THE

FUTURE

OF THE

3D PRINTING

CULTURE

Raymond R. A. Burke

Infovore, geek, Batman cosplayer, and wannabe Iceland explorer - Raymond Burke is a British-born author. His background includes a teenaged life in Canada and the US, his twenties in the British Army as an aircraft technician, his thirties as a mature archaeology student with BSc and MSc degrees from the Institute of Archaeology, University College London, and from his forties as a sci-fi author and freelance writer.

Raymond cunningly lives without a fridge, satellite TV, iPods, and he also can't drive. And while he has taken up 3D printing, he's a self-confessed 21st-century caveman - and loves it!

Through all, he has been a keen writer. He lives in London.

Acknowledgments

To my constant supporters John Mcmillan, Mark Emsley, Nigel Livingstone, Jenny Stripe, and Lori Buttermark. To my fellow writers David P. Perlmutter, Jon-Jon Jones, Stephen Marriott, Anne John-Ligali, Soulla Christodoulou, and Elisa Gianoncelli.

To Patrick M. Powers and the Founders Nation team and fellow entrepreneurs; Jay Jacobson, Hannah Jacobson, and Nour Youssef at Book Award Pro; and Thomas Anderson at Literary Titan.

I would especially like to thank Adam Bute of *Yes, That's 3D Printed* and Ravi Toor and Lakshit Kumar at Filamentive for their instructive feedback.

Book formatting and cover by Saheran Shoukat.

And a huge thanks to London Software Training, who started me out on the physical journey of the 3D printer adding to my intellectual curiosity about the industry and hopefully, its future culture.

To

YOU

the 3D Printers of the World

You're genius makers.

Experimenters of the future.

Create the 3D printing lifestyle you want.

Let's change the world one 3D printer at a time.

Design, code, print, repeat.

Just don't think out of the box,

create and open new boxes.

Let's get started...

Table of Contents

Introduction..11

Chapter One: The Culture of 3D Printing................................15

Chapter Two: The Household Appliance of the Future 21

Chapter Three: Furniture: The 3D Printing Station 33

Chapter Four: The 3D Printing Travel Station 39

Chapter Five: 3D Printing Fashion...................................45

Chapter Six: 3D Printing Recycling..................................49

Chapter Seven: 3D Printing Media 73

Chapter Eight: 3D Printing Events & Networking 103

Chapter Nine: Exponential 3D Printing 127

Chapter Ten: Environmental 3D Printing..............................157

Chapter Eleven: 3D Printing X-Prize.................................177

Chapter Twelve: The Cultural Future of 3D Printing 191

Chapter Thirteen: The Story of 3djacent Solutions...................209

Appendix 261

References 271

Introduction

I first came across the notion of 3D Printing in Douglas Mulhall's 2002 book *Our Molecular Future – How Nanotechnology, Robotics, Genetics, and Artificial Intelligence Will Transform Our World*. As I went back to the book after almost twenty years to refresh my memory for notes, I noticed something strange as I looked through the text and index.

Neither the words or terms 3D printing nor Additive Manufacturing were listed. Now I knew I had read it somewhere in the book as I could see the pictures of 3D-printed objects. When I searched again, the terms desktop digital fabricator, desktop fabricator, desktop factory, and desktop manufacturing were used. At that time, even twenty years after its invention, 3D printing was still in its infancy with no clear identity. And it still seems that way today despite the enormous uptake of the technology.

Mulhall, (and Neil Gershenfeld before him in 1999 with *When Things Start to Think*), described desktop fabrication as the time when corporations will become flat and decentralized as people learn to manufacture products for themselves. The service and brand industry would be revolutionized as people became their own service providers and brands. Humanity would be liberated from a consumerist world able to produce any number of miraculous life-changing applications.

And yet, another twenty years on and 3D printing has not matured enough to be mainstream. It's not an individual tool of liberation in most minds. When I told people around me about my 3D printing direction, they had either never heard of it or they vaguely knew you made things from 'cheap' plastic. It is not a technology high on people's agenda, yet has the potential to change each and every one of their lives.

So, this is the theme of my book. I'm in my fifties and have heard about 3D printing for the past twenty years and took it up only a few years ago. While I'm not a great designer, I have my own printer. I realized quickly that apart from the printer and associated parts and supplies, there was nothing else. I wanted something bespoke to place my printer on; a specific 3D printer-associated table. There was nothing. The 3D printer was an isolated technology. There was no culture springing from it, no links to the non-3D printing world to draw the uninitiated in.

Gaming, sports, music, geekdom. All have their cultures. You recognize it from their physical tech, artifacts, clothing, brands, symbols, advertising, etc. 3D printing does not have this. It's invisible beyond industry boundaries. Is this because it is a jealously guarded technology? Or because traditional industries see 3D printing as a threat and suppress it? Or because there is a lack of proactiveness in spreading the 3D printing word?

This is why I created my company 3djacent Solutions. I wanted to kickstart that culture within the 3D printing community. I wanted to help discuss, innovate, and launch the 3D printing world to the rest of the world with a view to realizing the vision mapped out by those like Mulhall, Gershenfeld, and others. Forty years after the invention of the 3D printer, I want 3D printing to be that household appliance everyone can have by the time of its fiftieth anniversary.

Impossible goal? We'll see. As well as my thoughts on the 3D printing industry (in which I have no job affiliation or vested financial interests in the featured companies or individuals), its culture, and my attempt to map out the next decade of 3D printing, this book is also the story of my journey to create my own 3D printing company to create that future.

The next few chapters will lay the path for the case for the 3D printing culture and explore the technology to be involved with this endeavor. I will be discussing and assessing the cultural merits of 3D printing in sectors such as: furniture, fashion, recycling, the media, events, and networking, with exponential technologies, the environment, prizes, and then a deeper look into 3D printing's cultural future. And in some cases, I'll offer solutions to bridge the cultural gap. I had also sought to contact a diverse range of people connected with the ideas I had developed and from their inspiration. Some didn't respond for whatever reason, and I get that. My ideas and solutions may seem naive, simplistic, complex, or impossible, but this book is also an open letter, the start of a dialogue, an issued challenge to 3D printers to be part of the cultural revolution.

The last chapter of the book covers my drive and journey to create the 3D printing culture. Why include my company story? Firstly, it's about the business culture of 3D printing and how I envisioned integrating that future through my company. Secondly, I wanted 3djacent to be a founder or cornerstone in the evolution of the movement. My successes, and lack thereof, led directly to this book. Most of the story is told in journal format and follows my early philosophical ideas on the subject. And of course, since 3djacent was the only type of 'cultural' 3D printing company I knew of, it was my only reference point for examples. Perhaps future editions of the book will include more successful companies and endeavors. Thirdly, I hope this also sheds light on my pitfalls and highlights and whether the 3D printing culture is actually a real thing or can be manifested. I leave that to you to decide.

So, you can learn from my story, maybe be inspired, and build upon, adapt, and improve on my ideas. It's all laid out for you. But lastly, above all, 3D printing isn't an isolated industry or a futuristic toy. It will be an indispensable part of our cultural futures. Therefore, along with the technical aspects in the book, and the speculative views of the future, there is also the business cultural story. My story. And likely yours as well. I hope it's a legacy we can leave in good hands.

CHAPTER ONE

The Culture of 3D Printing

This isn't another book about the future of 3D printing, *per se*, espousing the technology and changes to the world it will bring, though it will encompass part of that vision. This also isn't a typical reference book or a researched scientific work about 3D printing. It is intended for different readers with varying interests and includes detailed technological explanations as well as some blue-sky thinking. It is about 3D printing as a lifestyle. It is about 3D printing with cultural implications. There are present-day and projected cultural views on 3D printing and what the future could be like for the millions of people who embrace 3D printing and have it integrated into everyday life.

What is a 3D printing culture? Well, that is what I will be asking, exploring, and celebrating. Besides a 3D printer, the filament, and tools necessary for the 3D printing individual or company, what makes a 3D printing community? What is its culture?

In general terms, a culture is the continued beliefs, values, and knowledge of a society passed down to others. That includes its activities, art, and social pursuits. This isn't about word semantics over the meaning of culture. Each niche of society has a culture. People often say, "this is the culture of our company", our sports club, etc. Whether they state it as their ethos, the attitude, or the rules of a particular company, group, community, or country, they are talking about their culture. Culture is a set way of doing things, its own rules, vocabulary, and material features. You can positively state that art and music have their own cultural aspects; as do sports, gamers, geeks, and even politics. The list is endless. I will be exploring many cultural aspects in relation to 3D printing's interaction with them. The overlap between ideas of the future, 3D printing, and culture presents a unified overarching theme.

So, as well as asking what would a 3D printing culture entail, I will also ask why does this matter?

To begin with, I'm not an engineer or tech guy, a designer, or a professional 3D printer. I'm a writer and an amateur printer. I print objects for myself, friends, and family. However, I am a passionate advocate looking forward to the benefits 3D printing can provide in the future and what is required to see the fruition of that vision. I'm also not the typical profile of a 3D printer, being a black man in his fifties. But a cultural approach for me isn't just about my ethnicity or inspiring more BAME (Black, Asian, and Minority Ethnic) people or more women to get involved in 3D printing, which I hope this book does achieve. Culture involves everyone from your home, school, work, and leisure activities. So, how will this cultural journey start? I will elaborate a little on my 3D printing journey.

Back in May 2019, I started a 3D printing course, something I had always wanted to do. And I loved it. I even bought myself a 3D printer in July of that year. But what I found soon after was that there was not much in the way of a supporting structure around 3D printing, especially for home printers. I had imagined myself as a budding 3D printing entrepreneur pumping out trinkets to sell. But I came across two issues:

1. Most people and small companies are working on a print-and-sell basis. I call it the early Amazon model where they're just products to sell. That's good, but very competitive and possibly short-lived and uneconomical for some.

2. The printer was more or less a stand-alone technology within the home. I was looking for furniture and equipment to enhance my printing experience, but found the market lacking.

Everyone I asked in online 3D printing groups stated they either just got an Ikea table or made something themselves. There was no current market for 3D printer furniture.

I found that quite unbelievable and disappointing. But I had to ask myself, was that because such a line of furniture was not required as makers could just use a coffee table or old workbench, or was it because such products had not caught on yet? I really believe it is the latter, and that there are whole business models of opportunities missing or going under the radar. 3D printing has the potential to change the lifestyles of homes, businesses, and whole communities if that potential can be realized. It opens up new opportunities for homes and businesses.

So, I thought I would form my own start-up: 3djacent Solutions Ltd. The name is a play on the word adjacent meaning next to or adjoining, with the '3' replacing the first 'a'. So 3djacent provides services and deals with issues and equipment adjacent to 3D printers and the industry. The 'd' in the 3djacent logo is a stylized triangle symbolizing a play or a start button. 3djacent was set to manage your 3D printing lifestyle.

Just like TVs, computers, and gaming hardware have their own furniture to enable a more satisfying experience, so should 3D printers. One of the goals for my start-up would be to furnish designs to enhance the 3D printing experience. I even sought out furniture designers and manufacturers to create furniture to house, store, transport, and protect 3D printers and accessories. Just like you, I think one should enjoy a dedicated 3D printing center and be in that environment. The initial idea was patterned on the Ikea business model, the bespoke 3D printer furniture can be factory-made with flat pack, mix 'n' match, or designed with your own options, or even 3D printed.

In conjunction with that is the technology around 3D printing. I think in the future 3D printers will be included within the hub of the home technology and design center, comprising the laptop, smart TV, home AI, and gaming sector, etc. It will be integrated with these systems, even allowing you to print using each device. I will elaborate further on these ideas in subsequent chapters on how to better incorporate current home technology with the printers to also benefit those devices.

I believe 3D printers will be at the forefront of the Internet of Things (IoT) with remote 3D printing forming part of smart homes and businesses. There are Wi-Fi-controlled 3D printers in association with AI for voice-controlled printers. Voice control is a growing sector and many companies are exploring methods to improve this technology. In hand with this is the growing application of remote viewing of printers. If you have a 3D printing farm, both voice control and remote viewing will be of great benefit to the business.

Moving away from the early Amazon merchandise model, the 3D printing industry will have to diversify offering more services like apps, integration with other home or business devices, and providing the environment in which to do so with bespoke furnishings.

So, one objective of 3djacent was to seek out like-minded individuals who shared the expanding 3D printing philosophy and incorporate them into the rapidly expanding industry. There are a multitude of adjacent possibilities available for 3D printing rising to an exponential rate in the future. And I want to help create that future now. It will be an interdisciplinary endeavor with multiple partnerships being formed to achieve this mission.

So, how will this 3D printing culture manifest itself? Well, as a former archaeology student, I was taught about the material culture of people and what we could learn from them. The same is true for 3D printing. For the industry to really thrive in the coming decades not only will businesses have to adopt an outgoing culture, but so will the domestic environment. It is these cultural aspects and connections that will make people interested in 3D printing, bring 3D printing into their everyday lives, and make 3D printing the cultural touchstone of the future.

Homes across the world will have 3D printers, not just as a work or hobby gadget, but as an integrated part of the household within a creation hub. And the way to introduce, interest, and sustain the domestic growth in 3D printing would be by cultivating a material culture around 3D printing.

We humans like collecting and talking about our stuff. We buy, sell, make, trade, and discard tons of material culture every year. Our homes reflect our personalities and habits, much of that through our material culture. In regard to 3D printing, much of its material culture is the machinery itself, its supplies, tools and spares, and the material it makes. But what the 3D printer makes is not always the material culture in and of itself.

What we need, as stated above, is the furniture, the complimentary accessories and equipment, the fashion and art to visualize the culture, and the advertising and marketing to bring it to the attention of the general public. This can be 3D printed, but for now, will be traditionally produced. The 3D printing culture lifestyle should come with its own accoutrements, and accessories like fashion, hardware, and adjacent spinoffs.

I have mentioned a 3D printing lifestyle, but what is this? Have you ever watched MasterChef or a similar cooking show where the contestant states they have been dreaming about what to cook, menus, and ingredients? Or listened to a successful sports star or singer discuss their training regimen? From dawn to dusk they are immersed in their lifestyle. We all have a lifestyle, maybe not as defined as a celebrity's, but we all have a way of life we desire. It's no different for a 3D printer user, whether a professional full-time 3D printer or an amateur building a side-hustle.

What I wanted to create and manage for 3D printers was a lifestyle solutions service; a sort of 3D printing concierge company, beyond a typical 3D printing as a service model. 3D printing is accelerating as a global interdisciplinary and multidisciplinary industry. 3D printing has applications in the medical, food, building, industrial, aerospace, education, and recycling industries. 3D printing is for the big STEM manufacturers, but there are also hobbyist and entrepreneurial opportunities for freelancers and crowd-funders. I wanted 3djacent to cross all these boundaries and offer that lifestyle solution; everything you do in life, but dedicated to 3D printing.

On my website, the basis of the lifestyle management and solutions plan would have started with the directory of services and industries, such as printing services and the industries mentioned above. Each directory would have appropriate companies signed up and listed providing a comprehensive database of 3D printing companies across each industry, globally. 3djacent would then provide further services for promotion and networking purposes. There would be web pages for:

- Classified ads for searching for and posting 3D printing jobs.
- Marketplace to sell 3D printing and associated products whether directly through the company or through affiliate programs.
- Courses - whether 3djacent courses or advertised by other companies.
- A comprehensive social media platform.
- A merchandising and shop page for 3djacent branded products like furniture, t-shirts, and office and marketing materials.

I was especially proud of trying to promote independent and freelance printers, plus those on crowd-funding sites trying to create new 3D printers and products. Support and investment would be critical for them. This was a chance to try and pull the 3D printing field together, to create a community, a safe space for all things 3D printing —the beginning of a 3D printing culture.

Yes, other companies and individuals may already do the above to some degree, but it is an internal-based strategy to sell more of their own products. They are competitors. 3djacent is open for all. I wanted to inspire and innovate beyond self-invested interests. And now, I want to invite you to join me and others in introducing everyone to the 3D printing culture and the benefits it will produce in the future.

CHAPTER TWO

The Household Appliance of the Future

The original title of this chapter included the word gadget, not appliance. But I realized as I was writing that the 3D printer was more than just a gadget and had to evolve into an appliance to pass the cultural sniff test. My ultimate goal is to see 3D printers in every household like a TV or microwave. We need to take the 3D printing industry beyond the gadget tag and create an indispensable appliance. By 2033, the nominal 50th anniversary of the invention of 3D printing by Chuck Hull, 3D printers will be sophisticated enough to be a plug & play appliance with apps and uploaded catalogs for household members to choose preloaded designs and products or to make their own. And just like TVs, game consoles, and phones, consumers would want more than one with different models for adults and children, and with available upgrades. The 3D printer of the future would be an investment like a smart TV, car, or house.

People may still ask what's the point of having a 3D printer? Why would I need my own personal one? People used to ask what's the use of a car when it first arrived. What's the purpose of a TV? Or a personal computer? Why have a smartphone or tablet? The detractors don't ask such questions anymore as their benefits became clear. Once those technologies found their cultural niche their use became indispensable.

But 3D printers are expensive, I hear you say. Remember the cost of the first mobile phone, digital TV, games console, or tablet? They were almost unfeasibly expensive, an *objet d'art* to dream about. Now look at them. We can't do without them and customers will pay anything for them. Most desktop 3D printers are comparable in cost to mobile phones, laptops, and game consoles, so they are quite affordable.

But how would I use a 3D printer for everyday life, you ask? I haven't taken a course or I'm not an engineer or designer. However, just as you would microwave a meal or use a laptop and not have to take a course to use it, so should you be able to buy a 3D printer in the not-too-distant future and casually use it. Just download what item you want from your 3D printing subscription service (like a catalog), add it to your print account whether on a laptop, mobile app or flashdrive, then like your microwave just press a button and print it up on your printer. Can't find something you want in your size, color, or style or it's out of stock then make it yourself or design and order it yourself. You'll be 3D printing songs as art (see Chapter Nine) and art from AI creations. 3D printing will be your cultural hub.

Will people lose jobs to the 3D printing industry? Most likely; the consequences of which I discuss in Chapter Twelve. However, as with other industries, people and businesses will adapt. Your 3D printer could make you money — selling products, selling waste filament, and selling time for others to use it. Businesses can rent out machines. People can invest in a 3D printing company or let a machine do the work for them and make money just like investing in real estate or renting out cars would.

As Peter H. Diamandis and Steven Kotler have observed in their trilogy of works (*Abundance* (2012), *Bold* (2015), *The Future is Faster Than We Think* (2020)) we are now facing a world of the 6Ds where digitized technology goes through a deceptive phase before becoming disruptive to traditional industries and dematerializing physical objects, thereby demonetizing high manufacturing costs, and democratizing the use of such tech. 3D printers are fully embedded in this cycle as described above.

The main issue with home use would be the User Interface. How to make the 3D printer a plug & play appliance, whether producing both products or food? From my archaeological background, perhaps this would be a task suited for an urban anthropologist to study in conjunction with an engineer. Also, creating compatible software and hardware to link 3D printers to a smart home or smart Building Management System (BMS) system would also be required.

How humans react to and with new technology is a complex affair. Smartphones are intuitive and tactile. 3D printing for the non-technically inclined will have to embrace this approach with in-situ or remote interfaces. Currently, you have to go through the design/download a model phase, convert to an STL file, send to a slicer, transfer that file as a G-code to the printer, set the print parameters, and then print. It can be a tedious process, especially setting the printer options.

A one-stop-shop 3D printing appliance will have to be limited to final push button/voice control activation, just like a microwave, smart TV, or online ordering. The best option would be using a printer-mounted app-loaded touch screen as with mobiles and tablets. As part of the printer or by having Wi-Fi-connected devices (e.g. mobiles, tablets, cameras, etc.) networked to the printer, touch screens could offer easy accessibility, monitoring, and streamlining processes. Some 3D printer users already use some of these functions now. With the touch screen or networked system, you could upload/download print files to/from a cloud database for storage/retrieval straight to the printer, by-passing the current multi-stage process, which requires extra technical and engineering capabilities. That should be the objective of the all-in-one printer. Easy. Or easier said than done? My hunch is that companies trying to build such a printer are concentrating on the technology and the machine, rather than focusing on the most important thing—the user. Once they start developing for the user experience, or as discussed below on problem solving, then we will start to see successful, practical domestic 3D printer appliances.

3D printers should be included within the hub of the home technology and design center, comprising the laptop, smart TV, AI apps, and gaming sector, etc. It will be integrated with these systems allowing you to print using each device. See an object on TV or in an online game you like, just interface with your AI and ask your smart TV or gaming console to select/scan the object from the screen and send the file to your 3D printer. Better yet, in the future you'd be able to preview it on a holographic platform before printing to your specification.

The goal of Microsoft is to have a laptop in every household. I would posit that 3D printers are more valuable than laptops in a home as 3D printers could help create laptops and components to a specific degree. If not physically in every home, then access would be granted from dedicated 3D print shops or online retail units to print from a catalog or from your own design. While this is a service now, it is not a widely utilized service by domestic sources. In such times, with the Covid-19 pandemic era upon us, 3D printing shops should be in use more so than Amazon.

3D printer technology is still not fully mature and is still evolving, just like mobile phones—from brick to smartphones to wearable tech; and analog TV to digital screens to streaming. 3D printers will at least start off in homes as a smart product, maybe as part of the IoT, Metaverse, 6G forthcoming revolution discussed in Chapter Eight. So, with the exponential rise of 3D printing, why don't we see 3D printers being sold in high street stores and malls like other household appliances?

The 3D Printing Physical Retail Explosion?

In the first edition of this book, I had lamented the fact that 3D printers were not being sold in physical stores. I wanted to see 3D printers moving from online to physical stores for customers to see 3D printers in action like any other appliance. Compared with current technology and appliances, such as smart TVs, mobile phones, laptops/computers, microwaves, etc., 3D printers are mostly advertised online. Is that because of a perceived or genuine cultural perception of 3D printing as a high-concept abstract mechanism comprehended and utilized by tech 'elites' online? Or is online the new offline for new technology? Whatever the case, jumping 3D printers to a physical retail setting could elevate its publicity and sales, before reverting to an online or split online/offline advertising model when 3D printing is more available and user-friendly just like today's technology.

So, imagine my surprise when I was on a temp job in south London when I noticed a 3D printing store nearby on Google Maps. I hadn't realized there were actual stores trading in 3D printing wares with showrooms for customers. I then searched for other such stores and picked up on a 2016 article by 3D Printing Industry who at the time listed 33 stores in 17 countries. I contacted a few of them to learn more about their company and their experiences as a physical retailer and in attracting general interest alongside those who were already 3D printing aware. I reasoned that having such shops would greatly increase interest and adoption of 3D printing in domestic settings further pushing the technology.

One company, that didn't want to be named, did contact me. They once had a physical store and showroom, but then Covid hit. Demand died as physical visits decreased and online venues like YouTube created greater content and interest in printers. They added that once the shine of the new technology wore off, customers were less inclined to experience the in-person tactileness of the printer and were happy to do their research and buy online. The company could not justify the overheads of continuing with their physical store, so it was closed down and the showroom is also used for warehousing purposes. They are an online retailer again. I was more than a little surprised by this as I would personally have preferred seeing and touching a printer (as I had during my course) before buying it. But I guess with higher quality videos doing the rounds online, potential 3D printing customers can see how a printer works, what they want in a printer, learn 3D printing online, and buy the printer of their choice online. Perhaps 3D printing is more suited for online merchandising.

There are surviving pioneering brick-and-mortar 3D printing retailers. Between them, besides selling and renting 3D printers and materials, they also offer on-demand print and design services, bespoke and/or free software, 3D printing classes, on-site support and training, maintenance/repair services, and community events. Some have the normal walk-in process, but others are by appointment only.

I debated including the list below of such stores, as it would soon be outdated, just like the 3D Printing Industry article with shop closures and companies going into administration. But, I decided to add it to show you the companies that tried (as I had with my online company) and reveal the burgeoning depth and breadth of the 3D printing physical store industry. So, check out the stores, support them, but also check before visiting that the store is still in business and when you can visit.

- iMakr – With stores in New York, London, France, and Denmark.
- Get Printing 3D – Dayton, Ohio and Evanston, Illinois (as MakeXYZ).
- Matterhackers in Lake Forest, California.
- Imagine That 3D in Salt Lake City.
- Ideaz3D located in Mexico City, Mexico.
- 3D Printing Dublin.
- Cubeek in Elancourt, France.
- Unic-3D in Ixelles, Belgium.
- Tresdenou located in Barcelona, Spain.
- Repro 3D in Valencia, Spain.
- 3DiTALY with several stores throughout Italy, headquartered in Rome.
- 3Durak located in Istanbul, Turkey.
- 3Dee Vienna in Austria and 3Dee Budapest in Hungary.
- Beta2shape GmbH in Munich, Germany.
- Maker Point in several Netherlands locations.
- Zortrax Stores in Warsaw and Krakow, Poland.
- 3D Factory located in Tel Aviv Jaffa, Israel.

- 3D Printer Super Store located in Breakwater, Victoria, Australia.
- Raspberry Pi store in Cambridge and Leeds, UK.

This list isn't exhaustive by a long shot and I hope more stores will be on the way. But some of them will have to be more accessible with more support and services offered so they are open beyond just appointment occasions. Further, beyond just selling the machines and accessories, the whole cultural gamut of 3D printing can also be shown with merchandising for furniture, fashion, food, recycling and energy considerations, internet of things (IoT) capabilities, media interests, and academic and business opportunities. These would be artifacts of the 3D printing culture.

It's often stated that products aren't things people buy; products solve a problem for the customer. Shops sell problem solvers. Is 3D printing in that equation? Physical 3D printing stores could help realize the potential of what opportunities culture could bring to 3D printing and vice versa. Shops are the cultural places, cauldrons of technology and people, where customer experiences intersect with what problems 3D printing could solve. It will be these physical retailers that will lead to a better understanding, appreciation, and development of how to create the one-stop-shop 3D printing appliance.

Perhaps there should be tests and limited trials with 3D printers salted into other appliance stores alongside microwaves, laptops, and fridge freezers with a dedicated salesperson showcasing the features and advantages. Or there could be partnerships between traditional manufacturing bases and 3D printing retailers to share space and co-advertise. Once 3D printers become viable domestic problem solvers their cultural value will increase, demanding more visibility in stores. Casual shopping or window shopping may lead to 3D printers being discovered organically and lead to greater interest in 3D printing. Even if it doesn't lead to a sale at least a visual understanding of what a 3D printer is and what it does can be gained.

I guess the number of 3D printing physical retail stores will depend on the demand for 3D printing for the household or business. However, retail shops with 3D printers will quickly demystify the technology giving customers a real visual, tactile, and intuitive experience. It's how we humans work with technology. And what better way to interact with technology than with our cooking abilities? The kitchen will be upgraded with a new cooking appliance based around 3D printing.

The New Food Revolution

For foodies it would be a new culinary age. Currently, there are chocolate, confectionery, faux-fish, and pasta being 3D printed. And the rise in non-meat protein products would introduce easily printable meals. Love a chocolate gateau from a TV chef's menu? No problem. Select and download the ingredients list from your smart TV and send the file to your food printer. You would load in your ingredients or order them delivered by a 3D-printed drone. Once loaded, the printer could be set to print the cake as seen on TV or to add your own little touches. Then press the start button and print up your meal. Most likely, printers will be used in conjunction with a grill or heater. Also, the furniture around a 3D printer/cooker may change, with kitchen countertops altered to accommodate the provision of 3D printer-friendly utensils, materials, food dispensers, and food printing stock whether in pellet, filament, powder, gel, or strip form. Just print up whatever food you need, heat if required, then serve.

We could 3D print more than just meat substitutes with foods from the other main groups such as plant-based and fruits, carbohydrates, other proteins, dairy and alternatives, and even create oils and spreads in gel or solid forms. These 3D printed foods could provide anti-allergy, anti-phobia, and novel options with new flavors, shapes, texture combinations, and further research opportunities.

In Japan, Masaru Kawakami, an associate professor at Yamagata University, is researching and creating a 3D printer for a range of tasty soft foods for senior citizens. The menu would include meat, fish, proteins, or vegetables paste in gel or ink form. He hopes this will allow seniors to be able to chew and swallow food better, and also reduce the caregiving load. A food 3D printer would allow for a personalized diet, adding preferred taste, firmness, and texture options. The printer would work with an inserted food gel 'cooking' in the printer in around five minutes.

In a country with an aging population, Japan's caregiver facilities are overstretched with Japan's Ministry of Health, Labor and Welfare calculating there will be a shortage of 220,000 caregivers in 2023 rising to 690,000 in 2040. Elderly patients with chewing or swallowing problems will face ever-increasing services. Kawakami admits 3D printed food isn't a replacement for human-made food, but he does see the need for better alternatives to the current food for seniors. The main issue is the lack of commercialization of the food 3D printer. While there are other factors such as cost with Kawakami's printer potentially costing several million yen per machine, researchers also need to "simplify the time-consuming process of preparing ingredients before they are poured into the machine."

Food 3D printers will be very useful before they become part of the larger household creation hub. Food is such an important cultural aspect to humans and many detractors will reject the notion of 3D printer cooking. However, as with technological changes that have changed the face (and taste) of cooking over the millennia, 3D printing will be a feature in many kitchens, professional and/or domestic, in the very near future. These types of 3D printers could perhaps replace toasters, microwaves, grills, and sandwich/toastie makers that clutter countertops and even take the physical place of the cooker/oven to save space and costs, especially on gas and electric bills.

The Cultural Threshold

As I have done and will continue to state throughout, the cultural future of 3D printing is on the way, but I'm ahead of the times. I say this with confidence, based on other technologies which started out as technology for other industries and then filtered into the public consciousness as the technology became commercialized and culturally relevant. Think of space applications like memory foam, wireless headsets, scratch-resistant lenses for glasses, LED lighting, portable cordless vacuums, and more. Think of military technology that finally made its way into our cultural lives, like the Internet, microwave ovens, GPS, duct tape and superglue, cargo pants, and a lot more. We don't think about their origins anymore. These inventions and applications became a part of our lives because someone enterprising realized the cultural value of the product and adapted them to solve a problem for us civilians. While 3D printing does have a domestic presence, it is still not a fully formed, cultural appliance as ubiquitous as all the technologies mentioned above. It hasn't passed that cultural threshold test, yet. It will. And it will rely more on our changing cultural mindset than on the technology itself.

This may involve the changing of the 3D printing hardware; its current physical form to fit the domestic setting, the materials it uses to print which could include safe dual-use materials that are also edible (like hemp), or having easily exchangeable parts for easy repairs and printing multiple materials at once. Whatever the change, 3D printers need to be optimized for domestic usage.

In fact, 3D printing may only become an indispensable appliance in the wake of a new cultural invention, maybe something not even realized yet. Very likely it will involve the application of AI as an interface. Perhaps it will be precipitated by another pandemic where 3D printing devices and files are shared globally for mutual benefit, or a new metaverse gaming feature specifically requiring a 3D printer, or a nifty food dispenser, or a cultural phenomenon not dreamed of yet. Whatever the precursor is, when it occurs it will enable the 3D printer to cross that cultural threshold and truly be a part of our lives.

With 3D printing technology improving and costs decreasing, this creates a virtuous circle of tech advancement and lowering costs. This makes 3D printing technology more attractive for the domestic appliance industry. At some point, 3D printers will be advertised on TV like other household appliances, whether sold in physical shops or online. Then we'll have kids yelling at their parents that they want a 3D printer for Christmas. But first, I believe you need an appropriate platform from which to work. It's time the 3D printer had its own furniture set.

CHAPTER THREE

Furniture: The 3D Printing Station

Why furniture? Or specifically, why furniture for 3D printers? Furniture—let's call them adjacent physical assets—gives a field of industry or entertainment viability and visibility, a niche of existence beyond the machine. Furniture would help grow the 3D printing industry, help normalize the use of 3D printers for domestic use, and create an underlying base from which to grow the domestic and even commercial industries.

Think about it, it doesn't matter what kind of TV, desktop, or laptop you have there are a myriad of TV stands or computer desk options for them. So, why not for a 3D printer? I wanted to create a 3D printing center, an ergonomically engineered area for the crafting of your goals. Is that too much to ask for?

Even before 2020, when I had volunteered for 3dcrowdUK to 3D print faceshield parts during the Covid-19 Pandemic, I quickly realized my coffee table was not enough to print faceshield parts, clean them, and have my meals on the same surface. The logistical process was nuts and I desired a proper workstation setup for the printer, the spare filament spools, tools, and printed parts. And most of all for cleaning as I was spraying the printer and printed parts with alcohol sanitizer.

Initially, I envisioned a metal contraption for sturdiness and cleanability, lightweight, and easy to assemble. Wood could be used like an IKEA flat pack, but metal would suit the job better. But try as I might, I could not find what I wanted amongst the computer desks, coffee tables, work benches, trolleys, TV stands, etc. Those were dedicated to their own cultural niches. 3D printers had nothing, except for make-do, self-made contraptions. In fact, I had asked

Facebook groups about this and most stated they just got a simple Ikea 'Lack' table. That wasn't good enough for me.

So, what spurred me on? I had attended an entrepreneurs' Meetup. It was inspiring listening to people's stories, networking, not being sold to, but making connections. It was the first time I heard about digital nomads. What the hell was that? I didn't know at the time, but I wanted to be one. But, anyway, I mooted my idea about furniture for 3D printing to others. I got talking to someone in September 2019 about getting a CTO for my company and who himself seemed interested. The conversation progressed as I told him about my plans for creating bespoke furnishings for 3D printers. And after our initial conversations, he then came back with:

"Hi Ray, I took the time to read, and your project seems crazy?!"

Now, I'll admit that most of his reservations were because he wasn't in the 3D printing industry so he couldn't really bring any skills. In his words, he was "really far from this world." So, I had to look elsewhere for a CTO, which didn't come to fruition. As I replied to him, I noted that I was not trying to re-invent 3D printing, but I could see the possibilities in my pursuit.

But it was when he had told me that my furniture idea was crazy, I lapped that up! There's nothing better an entrepreneur wants to hear than his idea is crazy. It means others haven't seen or heard of it before. It was new, nascent, and ambitious. No one else has thought of it or was doing it, at least not on a large, public scale. It wasn't impossible. And I was going to do it.

So, what exactly was I looking to build? Well, I wasn't too sure myself, so I put out ads on freelancer sites like Upwork and Freelancer, like below:

I'm looking for a furniture and accessories maker, specifically making bespoke tables/workbenches (with adjustable shelving) and cabinet compartments for storage of tools, spares trays and racks, drawers, and shelves.

The material can be wood, metal, or composite materials with glass top or frontage. Further options to make some of the furniture lines flat pack. The furniture lines are for devices so some knowledge of computer and printer furniture designs would be great.

There should also be the option for future production line capabilities, if you don't have that facility already.

At the moment, I am looking for a prototype/proof of concept and collaboration for a much bigger project, so some entrepreneurial spirit would be ideal. Please do drop me a line if you're interested.

I got some replies, but none of them seemed right as they mostly promised they could do the job cheaper than anyone else, which wasn't the point. So, I came up with a better spec as detailed below.

In thinking up the 3D printing station, I had to wonder why it had not been thought of before. Was it ahead of its time? Was it a product that was really needed? Did it solve 3D printers' problems? But I had to test the waters. Whether the station was a successful product or not, at least hopefully 3djacent would be labeled as the first company to have launched the first mass-produced furniture item for 3D printers — my legacy.

Then, in December 2021, my heart skipped a beat as I saw that Ikea had entered the 3D printing market. I thought, 'great, they're making furniture for 3D printers now'. But no, Ikea had produced a range of 3D printed house accessories and decorations. You'd think they go for the obvious and manufacture a complete line of 3D-printed flat-pack furniture. That was the objective of 3djacent. But no, since 2019, Ikea has worked with 3D printing companies and 3D printing hack processes have been used to create products as slip-ons or add-ons for disabled people to use Ikea products.

3D Printer workstation design spec

PROJECT NAME	Temporarily designated as 3D Printer Workstation.
BACKGROUND	After buying a 3D printer, we realised there was no dedicated 3D printer furniture to place the printer on and other creators had fashioned makeshift platforms or made do with less optimal platforms. We need to create such a work platform for 3D Printer users to: 1. enable users to work properly with specific equipment designed for them. 2. help users manage their 3D printing lifestyle by providing a physical environment to grow their business.
OBJECTIVE	To create a dedicated work platform for 3D printer users which ideally suits 3D printer users' needs and affordability.
TARGET AUDIENCE	3D Printer users all ages and abilities.
KEY MESSAGES	This is the first 3D printer furniture piece of its kind.
LOOK & FEEL - FEATURES	Flat pack assembly.

	1. Table/work bench
	W/L - 50cm x 50cm (square) or 50cm x 70cm (rectangle)
	Height adjustable – 40cm (coffee table height) to 75cm (desk top height) - crank and lock mechanism or bolt and nuts).
	2. Frame work
	Metal, 4 legs ending in braked wheels, cross supports (shelving?), metal tray top supporting surface.
	3. Surface
	Wood top (perhaps removable fitted board to allow work on metal tray surface), protective covering for working on, cutting, cleaning, scraping, and modelling.
	4. USP
	Since there will be products similar to this we will have to differentiate our station from other normal tables/benches/platforms with ‚premium' or unique points. What would a printer user want?
	    1. Printer cover to avoid fumes and dust
	    2. Optional pull-out tray for laptop, tools
	    3. Spool rack.

MANDATORY INCLUSIONS	Has to have unique/specific functions or features to delineate work station as a 3D printer user's furniture.

So, in December 2022, I contacted Ikea, firstly to get their feedback/opinion on my furniture idea and secondly, to submit my ideas and design proposals as outlined above, with a crude mock-up of what I envisioned. Later in the month, I followed up on this with Stanley Black + Decker with the same details proposing to SBD that they could start a line of 3D printing work benches and other associated furniture. Between Ikea and SBD, two of the world's largest furniture designers and manufacturers, surely, I would be able to not just get a response, but also spark some form of action from them and kickstart the 3D printing furniture industry. But apart from a form letter reply, there has not been any follow-up from them.

And then in December 2023, I saw that 3D Systems, the first 3D printing company in 1986, had also started producing household furniture using their proprietary pellet extrusion printers. Their process eliminated petroleum-based plastics with their harmful carbon emissions and waste. Instead, they used recycled sawdust, bio-resins, and other non-toxic materials and plant-based waste to create 3D printable materials and products. While that is laudable, I still think even 3D Systems is missing the trick in creating furniture for their 3D printing consumers.

Offices, whether in workplaces or at home, have been designed for the typewriter and paper-age. Laptops and computers have seen an evolution in their shape, size, and style; plus chairs, desks, and associated equipment such as printers and desk lamps have also undergone changes to accommodate their shape and size and the habits of their human controllers. However, the 3D printer is now expected to stand and work on the footprint of the previous technological generation. It's not enough to make do. We spend our lives at desks sitting on inadequate chairs for the job most times. Sure, traditional manufacturing can solve these ergonomic and stylistic issues, but they haven't. So, let's up our 3D printing game and design and print the furniture that 3D printers can use to comfortably and safely carry out their trade and hobby.

Before the 3D printer becomes an integrated appliance in the household, business, or school without requiring specific furniture, printing platforms will have to be offered by tables and workbenches for standalone 3D printers. There would also be a need for storage units or 3D printer-friendly shelving for when the printer is not in use. My proposed printing station would also serve as a storage unit, its top doubling as a work surface depending on its size, with optional pull-out shelving sections. So, be mindful of the space and type of future bespoke 3D printing furniture that will inevitably be required in the living room, den, home office, or kitchen areas.

In a slight diversion from furniture, there are 3D printed houses springing up around the world whether single domiciles or large estates ranging from Texas, Dubai, the Netherlands, Germany, and the Czech Republic for starters. And yet, we are barely 3D printing traditional furniture, let alone furniture for 3D printing. Yes, traditional manufacturing and the building of houses and furniture may be less costly, but environmentally they are not sustainable industries. In Chapter Ten, I take a more detailed look into the environmental aspects of the 3D printing housing market. 3D printing will eventually enter the market offering hybridized building solutions or a fully 3D printed service. And then you can place your 3D printed furniture within.

But, what if I had to take my domestic printer elsewhere? What if I was moving home or giving an exhibition elsewhere where I had to move my printer off the premises? How would I safely protect, store, and transport the printer? And then I thought, why not also make a portable 3D printer station?

CHAPTER FOUR

The 3D Printing Travel Station

Hi,

I'm looking to have a transport case made for my 3D printer. Hard outer shell (metal), with telescopic pull handle, wheels and clasp locks. Inside would be hard foam insulation. Ideally, I would also like interior pockets for spool/spares/tools.

The dimensions of my printer are H: 62cm/24 ½", L: 46cm/18", Depth: 46cm/18".

I have attached a picture of the style I want. There may also be opportunities for modifications. I look forward to hearing from you.

That was my first approach to a manufacturer of transport cases. I had researched transport cases and came across three. One of the companies I initially contacted in July 2020 after seeing their ad for a travel case for a small 3D printer was the Zeepro Travel Case for Zim 3D Printer. It looked great, but after contacting them they told me that sales had been discontinued. I asked them why and other questions like:

1. Why were sales discontinued? Did it sell well?

2. Can other size cases be made - such as for Ender 3 to 5 series? 3. Did you have a manufacturer for the case I could contact to have others made?

4. How much did it cost to make? Is there a sample available to buy?

If you are willing to make and sell again, once my site is set up, I would like to arrange a meeting to see if we can sell the product on our site or advertise for you.

Such a case would have been ideal for demonstrations, trade shows, and travel protection.

The seller of the transport case did get back to me:

1. Whenever an item is listed as discontinued on our online store, it means that the manufacturer notified us they would no longer be providing said product.

2. For that, you would need to source a case manufacturer.

3. The manufacturer for that item is Zeepro, Inc. - they made it, we placed orders so that we could offer them for sale.

4. We do not have that information nor do we have samples.

So, for them it was the manufacturer who had discontinued sales (and maybe manufacturing), so I needed another source.
Luckily, when I went back to the handy freelance sites, one of the respondents contacted me to say he knew of a company called Flightcase Warehouse that may be able to supply such cases.

I contacted them and outlined what was required. They were positive with their response, quoted for the design prototype, which was reasonable, so I ordered one. I also had other ideas which I outlined to them turning the transport case into an actual printing surface; a dual-use 3D printing station:

One of the products I am looking to sell is a transport case/station. Basically, once the 3D printer is taken out of the case, the case top can be used as a platform for the printer (depending on if the handle does not get in the way). Then the interior can be converted into a shelf unit (2 shelves for the spools, tools, and accessories) with the shelves (part of the hard insulation?) sliding through slots in the insulation. We then have a mobile 3d print station. There may be other modifications that I can share later. My printer is an Ender 3 Pro, but other size printers could also be catered for with other cases. It would be a new industry product... I believe such a unique product would be very popular with 3d print enthusiasts and for events and promotions. Please do let me know if this venture would be amenable to you.

After a brief delay, my travel station arrived mid-September. It was larger than expected and not designed totally how I wanted it in regard to the interior due to the nature of the case, but it was very much in line with what I required. Plus, it was only the first design, an MVP (minimum viable product), with more adaptations to come with plans to set them up as a supplier. I quickly wrote a script for my video and print ad:

"Like me, do you have a problem moving your 3D printer around town or storing it and parts safely away? Yes, well, I'm super super excited today to show you our new product for your 3D printing lifestyle—our 3D Printing Travel Station.

The travel station is tailor-made for your mobile printing activities such as attending exhibitions and events, moving house or student accommodation or work spaces, or as a teaching aid keeping the contents safe and for use as a working platform—just plug in and you're ready to go on a hard level surface.

It's a durable, lightweight elegant metal and hexaboard casing with two tough wheels and hard rubber feet. We've got heavy-duty fittings with 3 grab handles and a telescopic pull handle for safe maneuverability, easy-to-open butterfly locks with padlock tugs. It's a luxury you can't afford to be without.

The exterior measures: H- 74cm W- 53cm L- 70cm – or in old change H- 29" W- 21" L- 27 1/2 „ inches. Let's have a look inside.

Top tip: save and use the delivery bubble-wrap for the interior of the case when in transit.

As you can see the interior is spacious with hard dense foam to protect your printer. We have a main compartment for the printer, in this case an Ender 3 Pro, and an upper compartment for at least 8 boxed 1kg spools, tools, spares, marketing materials, or for smaller prints.

Of course, the travel station would also be an ideal extra storage unit for your printer when not in use and for accessories, especially good for that unused corner space. You could even decorate and brand the exterior for your own marketing needs.

In the future we can produce different sizes or interior fittings, just let us know your needs or if you have any queries by contacting me. Production and delivery will take 5-6 weeks.

I am really looking forward to you owning one of these bespoke 3D printing travel stations, you'll love it."

And that's how the 3D printing travel station came about. Here, I had borrowed from the travel industry culture to make the 3D printer more mobile. I always remember reading about how long it took for suitcases to have wheels attached (after hundreds of years of traveling). It was such an obvious solution and another material culture evolutionary step. Sounds simple, but just like a packed holiday suitcase, try lugging around a printer from place to place then trying to find a suitable set-up surface to operate and you'll see that a combined mobile, convenient, safe, and secure print platform is the way to go. And when you start decorating your travel station with paint, stickers, and other designs it will also add another avenue for sales, marketing, and visibility. Let people know you're a 3D printing guru.

Sure, in the future we may be able to go from home, to school, to work and use other printers, like hot-desking (I'll coin the word hot-printing, here), but for now, we have our own personal printers. We may be traveling exhibitors, mobile students, printing demonstrators, and teacher-trainers, so the 3D printer will be on the move. So, we'll need a dedicated piece of furnishing for that and at the same time make it useful in its stationary state.

So, paraphrasing the justice system we now have the means, motive, and opportunity, to transport, protect, and carry on 3D printing on the go. Having 'invented' 3D printing furniture sets, I looked for other cultural ventures 3D printing could ally itself with. And the obvious choice was fashion.

CHAPTER FIVE

3D Printing Fashion

What do t-shirts have to do with 3D printing? As the most ubiquitous material cultural fashion statement, t-shirt designs, slogans, and logos can propel 3D printing into the public limelight.

However, if you look through most merchandising sites for 3D printing companies, what's on offer is often limited to self-branded products, including clothing. But the clothing doesn't sell 3D printing, it sells the company, which the general public may not be aware of.

In looking at the bigger picture for 3djacent, one strand of revenue and merchandising I seized upon to increase revenue and add to the visibility and marketing of the company besides stationery and office stuff was t-shirts. The t-shirt industry is a multi-billion dollar one with the custom-printed t-shirt market growing every year with over 2 billion sold every year. Every one of every age owns a t-shirt, mostly displaying some kind of message, brand, or logo on them whether they share that company's values or not. For those of you worried about commercialization, commodification, and sustainability, t-shirts can be printed on demand using digital technology, requiring less water and electricity to produce, and using more natural materials. And with wearable tech on the way, the t-shirt will last well into the next century.

In February 2021, I branched out into t-shirt making, creating slogans and designs with a 3D printing motif. Using slang and verbiage from 3D printing (like maker, fabber, stereolithography, etc.) was a fun and educational way to go so the non-3D printing public could see a new 3D printing culture emerging from beyond the machine. It creates interest and engagement with the greater public.

By September I had over thirty 3D printing shirt designs (and growing), including my signature Spoolhenge collection and Maker Nation motifs (which basically had the words within an outline of a country) and some humorous but rude messages like Lick My Nozzle, Nice Rack, and Snozzle My Nozzle (with appropriate 3D-printed related images). To increase understanding and visibility, many of the designs or slogans were incorporated or borrowed from already-established memes, visuals, and slogans (such as "My mom went to an additive manufacturing conference and all I got was this lousy t-shirt!"). It makes 3D printing more relatable and slashes the work needed to be completely original and inventive.

3D printing fashion lines add to the cultural world of 3D printers and more marketing visibility. As I have noted before, most other lifestyles (e.g. sports, gamers, geeks, films/books, luxury brands, etc) have their own fashion brands and styles. Adding your own logo and message will thereby enhance your brand. T-shirts are cool, simply designed, and such a visible marketing agent for anyone.

However, what also needs to happen is for the corporate world of 3D printing to realize 3D printing isn't about suits. So many tech and 3D printing conferences are still stuffed to the gills with be-suited sellers. For an exponential technology on show, people in suits will not grab the attention of younger Millennials, Gen-Z, or Gen Alphas. You need youth brands and techniques, social media savvy and swag, and a means to reveal your brand quickly and coolly, and a t-shirt or even a short-sleeve shirt is the way to go. It creates instant impact, a talking point, extra revenue streams, and mass exposure. Even if people do not understand or use 3D printing technology, a cool t-shirt will still demand attention. And that would just be the beginning. This doesn't have to stop with t-shirts as other apparel like hoodies, jackets/coats, caps/hats, and other visual outerwear could also carry a 3D printing message.

And this is even before actual 3D printing clothing is taken into account. At the moment, shoes are the most commonly 3D printed fashion product with personalized 3D printed insoles also available. Clothing printed from plastics exists, as do simple garments ranging from dresses, jackets, 3D printed knitwear, ties/bow ties, glasses, jewelry, watches, bags, and other accessories also on the uptick. But much of it is not physically branded or commonly known to be 3D printed. It's a missed opportunity for marketing and awareness of 3D printing. While such products are on the market, they are almost 'restricted' to online markets. Until we see products like these in bricks-and-mortar household-name high-street shops or in dedicated physical shops with only 3D printed stock, then the public is not going to know.

So wearable material culture depicting 3D printing culture is the way to go in attracting interest, increase the 3D printing brand, and gain acceptance in niche markets. The more visible 3D printing is on 'normative' items like clothing then the better to introduce further 3D printing culture, philosophies, and technologies to the masses.

Technologies will include wearable tech which can be implemented by 3D printing. Wearing a 3D printed t-shirt or jacket incorporating 3D printed wearable tech, such as 3D printed flexible solar panels or piezoelectric crystals, which charges your cell phone or smartwatch on the go or keeps you warm or cool will be a staple of the future. The clothing will most likely be printed with smart biodegradable materials to allow for much better wearability, recyclability, cleanliness, repair, and adjustable comfort.

And with the metaverse on the way, your future clothing could be your entry ticket. Your 3D printed glasses with embedded sensors and wireless connections (perhaps with an AI assistant) would enable access with a few blinks or read of your eye for identification and authorization. 3D printed wearable tech on your gloves, chest, and sleeves would also have security and identification features and allow you to control your virtual environment, including advertisements and posting social media updates. You could even 3D print in the metaverse, as I discuss in Chapter Nine. Your individual 3D-printed fashion becomes your gateway to the virtual world. Your fashion becomes technology. And 3D printing becomes fashion.

At the end of the day, for people to invest in you and your 3D printing venture is to invest in your material culture, which would include fashion, especially t-shirts. Humans are makers, traders, and owners of materials so use this benefit to the best of your ability.

CHAPTER SIX

3D Printing Recycling

Some of you may question recycling as a cultural process, but it certainly is. Recycling, in the wider environmental culture, reflects our current and future outlook on resource exploitation, management, and reclamation. Our practice of recycling as a habitual act tells a lot about us culturally and our efforts to help the environment. And within that environmental context, existing and experimental 3D printing recycling methods can be a large part of the process assisting in our cultural needs and aims in preventing further resource loss and to preserve the environment as much as we can.

Let's be honest, we 3D printers work with a lot of plastic, where a great amount gets wasted as discarded and failed prints, supports, unfinished spools, etc. According to B4Plastics (more about them further below):

> "global consumption of materials such as biomass, fossil fuels, metals and minerals is expected to double in the next 40 years, with harmful consequences for humans and the environment. The annual plastics production is close to 380 Mt and is expected to double by 2035 and even quadruple by 2050, making it essential to find new solutions for effective and efficient use of resources for their production." (https://b4plastics.com/projects/mycomaterials-2/)

Even more shocking is research showing that more than 75% of new plastics ever created has ended up as waste. Of that waste only around 9% has been recycled, 12% incinerated, with a whopping 60% (equating to 5 billion metric tons) disposed of in landfills or remaining in the natural environment. That's a lot of plastic contributing to environmental pollution that 3D printing can reclaim, recycle, and reuse. We need concerted efforts to keep 3D printing green and help alleviate plastic pollution problems.

Fortunately, one of the most common plastic filaments Polylactic Acid (PLA) is recyclable, biodegradable, and compostable to a degree and there are many rising companies ready to recycle for you. You'd be able to save money on buying new filament and recycling the spools as well.

For 3djacent, I had written my vision on 3D printing recycling and how it could be shared by others. I called it my 3D Printing Recycling Lab. I made a video in my local park at the end of August 2021, describing the venture, and put it on my YouTube channel:

> "Hi, I'm Ray from 3djacent Solutions. Let's talk more about the recycling lab, our collaborative 3D Printing Recycling Project.
>
> As I mentioned on the website we want to recycle, especially the plastic filament, but we find there are no practical local services to do so and the costs of the equipment to do so ourselves is somewhat out of our price range. We may want to recycle filament to make new filament or to create new products or even try to recycle other domestic plastics to make new products. But is this feasible as individuals? Most likely not.
>
> So, what we're seeking are great recycling ideas: if anyone wants to get together to buy and share recycling equipment; if anyone has time to drive vehicles for plastic waste collection, if someone has storage space, let's get together and make our own recycling business. Let's learn and earn for ourselves.
>
> For the more ambitious and more connected of you, this also goes for metals and ceramics recycling. I want to know if we can make 3D printers from recycled metal, or literally turn swords into ploughshares;

guns and knives into pens and tools. That is the power we have. There will be lots of and's, if's, and but's, but if we don't ask the question of the impossible then we will never know.

So, write blogs & articles, share your ideas and videos, and publicize your efforts in the recycling lab. Spread the word. And let's grow the lean green 3D printing recycling machine."

Great, huh? I didn't get the word out as much as I wanted, though I had lined up a potential supplier of large end-of-life plastic water cooler bottles. But where I have failed, others can take up the baton and create their own 3D printing recycling business. This would be another cultural factor as 3D printing will have an enormous impact on recycling various material types as discussed below. I have included some technical details so you can comprehend what is involved in the conception of and processes used in 3D printing recycling techniques. Perhaps you can then use, repeat, experiment with, and improve upon the research and work.

Recycling Plastic

Currently, the recycling procedure works by shredding waste filament in a machine or manually for the melting process before reshaping it in an extruder, a device with a hot end and nozzle which controls the filament diameter required. The new solid filament should then be cooled before being carefully spooled to avoid filament deformation.

The big issue for plastic recycling is the consistency of quality. Strength and thickness values in the recycled filaments have to match the original filament properties, so the waste plastic must have the same material properties (i.e. not mixed PLA, PETG—Polyethylene Terephthalate Glycol—or other materials, etc), shred size and thickness must be near identical, and the fluidity of the melted filament be even throughout. There are few recognized standard quality assurances for recycled plastic filaments, so that would be a place to start to have an industry standard established.

Another factor in recycling is cost. DIY homemade filament recycling can be manually exhausting and time-consuming while specialized recycling machines and systems can be quite costly for solo/small entrepreneurs if you opt for an extruder, a shredder, and a spooler or even an all-in-one system. Some systems offer a pelletizer option as well. An advantage to many of the recycling systems offered by companies is that they often have ready-made communities ready to share their knowledge and feedback on recycling products, services, set-up, and operation.

But, if you cannot afford your own equipment, there are services that offer you the opportunity to recycle your materials promoting a greener 3D printing industry. Some companies allow for plastic waste to be shipped to them or to be dropped at local drop boxes for collection. In return, for every kilogram you send, the company will offer the sender discounts towards future purchases of their recycled filaments. Perhaps, there will be an app for this, discussed in Chapter Seven.

In March 2020, I contacted a company, called Filamentive, that sold recycled filament and enquired about their recycling process and what services could be provided. In a personal communication, the Director, Ravi Toor, kindly responded:

Evening Ray,

Ravi here. Many thanks for your email.

Thank you for your kind words. AFAIK there is not a complete market solution available.

Some hobbyists have solved their own waste issues with creative solutions (e.g. melting down and moulding into a product of kind). At the higher end, one of our 3D printing service clients is upcycling their own waste using desktop extruders.

As a provider of filament made from recycled plastic, recycling waste/ failed 3D prints is definitely an aspiration. Many operational and logistic concerns exist in regard to receiving waste, in addition to the obvious challenge of quality control.

ICYMI - we recently wrote a blog post titled: Recycling Failed and Waste 3D Prints into Filament: Challenges.

Until such a solution is achieved, we will continue to be the sustainable choice in 3D printing by committing to:

1. Using recycled material (post-consumer and post-industrial) to produce our filament where possible.

2. Avoid the use of new, virgin polymers to reduce energy and demand for raw materials.

3. Utilise plant-based bioplastics when there is no recycled alternative.

4. Forming strategic partnerships with recycling companies to use their waste streams to produce filament.

5. Using 100% recyclable cardboard spools to further reduce waste and increase the recyclability of our products/packaging.

Whilst we do not have a waste management service in-place yet, we are actively working on it and our end-goal is to produce 100% recycled, high-quality 3D printing filament - but should this not be feasible we are exploring alternative End-of-Life (EOL) solutions, such as melting waste PLA into pellets, and also organic treatment via industrial composting facility.

I trust that is all helpful info, Ray - primarily our business is offering premium 3D printing materials made sustainably from recycled materials so if this is of immediate interest I'd be happy to send samples.

Beyond that, we are very keen to curate innovative ideas to promote circularity of materials within the 3DP and so it'd be great to forge those links as you say!

Best Regards,
Ravi Toor
Director

I found Ravi's information helpful and indicative of the 3D printing recycling industry. Filamentive are one of the pioneers in the market and I hope they will be able to be part of the complete market solution when it arrives.

It's a challenge I'll throw out to you and to the 3D printing companies looking to expand 3D printing services and increase your green credentials and revenue.

However, plastic is not the only 3D printing material requiring recycling solutions. Below, I explore and discuss recycling options for alternative materials and how 3D printing with recycled materials will have an impact on our cultural environment.

Recycling Metals

This process would be much more intensive than plastic recycling and a mission more fitting of a concerted professional endeavor. One company looking to address this matter is MolyWorks. They are taking scrap metal and melting and atomizing it into powder suitable for 3D printing purposes. Their goal is split between decreasing the use of virgin metal and also allowing for localized recycling units to establish a distributed manufacturing network. So instead of mining, processing, and transporting new metal, scrap metal can be recycled at a local center saving time, money, and energy, plus less pollution in the environment.

The versatility of the metal powder allows for greater usage and smarter print choices. The current MolyWorks recycling unit, though the size of a shipping container, can be mobile, cost-effective, and not limited in the types of metals that can be recycled efficiently.

Of course, like plastic recycling standards, metal recycling will have to have strict criteria during the process such as knowing the source and type of metal and/or alloy, its chemical properties, and how it will react to the recycling process. MolyWorks has also developed a QR code system for each of its manufactured parts as a way to ensure quality assurance and to future-proof its procedures.

MolyWorks envisions its metal recycling as a future service where customers pay a fee to have their metal scrapped and recycled into powder. So keep an eye out for their service and get inspired on how you can recycle your metal as well. This would make a great franchise service where motivated printers can become involved.

In December 2022, I contacted MolyWorks' Dave Neufer—VP of Strategy and Business Development, and Tony Des Marais—Product and Services Director to inquire if Molyworks would consider a franchising business model for a fully-enabled distributed manufacturing network. My idea would see industrial or larger print companies hosting MolyWorks mobile foundries, but also permit enterprising solo-preneurs, 3D printing farms, or small companies using metal to form the basis of a franchising model. Such business models could help not only transform the metal recycling field, but also encourage advances in the recycling process. If franchising out the mobile metal recycling foundries were an option, what would be the criteria for such an operation? And what other details would have to be considered?

I was disappointed, but not surprised, not to get a response on the matter, so let's take this as an open letter to the 3D printing industry and a challenge to further the metal recycling capability within the industry. With the amount of appliances and devices used worldwide made from virgin metal mined in unsustainable methods, the key to less costly, less heavy, and more customizable metal goods will be through 3D printing with recycled metals. It's an industry waiting to explode with huge potential for households, businesses, entrepreneurial ventures, academics, and even the space industry.

An example of the seismic shift recycled metal could create in the economy would be governments using 3D printers to mint physical legal metal currency—coins. Whether currency becomes fully digital or not in the future, there will still be a need for some form of coinage for some transactions and businesses. However, with 3D printing coins, government mints would be able to substantially downsize on the hardware required to mint coins, save energy, save costs, and even save on currency transportation costs as coins could be reprinted where required on official 3D printer mint premises dispersed across any city. Centralized mints would not be required.

3D printers could also add security features to coins traditional mints could not. Current mints use various measures to prevent counterfeiting such as varying edge-designs, layering metals, varying metal content, tagging metals, serial number scan on file data, and micro-engraving identification.

3D printing security measures could entail:

- Using recycled metal in specific quantities which are sourced, cataloged, and in a database.
- Placing microdots of recycled metal in virgin or recycled coins for identification.
- Creating regional-specific coinage for general distribution and greater security and identification.
- Creating unique designs within the coin body and/or upon the coin surface traditional mints could not do, such as twisting or patterning the metal.

- Using 4D or 5D printing techniques to make coins extra unique. In effect, creating smartcoins, which would react adversely to being copied or misused.

The benefits of using recycled metal are limitless even if not used in the 3D printing industry. But why waste the opportunity (and the metal) by not trying to create a more sustainable future from such a common cultural resource?

Recycling Ceramics

As the oldest known natural ceramic, clay can be used with other earthen materials, powders, and water to make a myriad of products. These items may be constructed from natural material and thus be considered biodegradable, but they can still take thousands of years to decompose.

But, due to ceramic being hard-wearing and highly heat-resistant, it takes an immense amount of heat to melt such materials so special processes are required to break the ceramic down. However, ceramic usage in 3D printers is growing and it has been predicted that by 2028 the 3D printing ceramics industry will generate up to $3.6 billion globally. So, finding ways to recycle ceramics will be essential.

Companies such as Coudre Studio in Spain and Fabrique Publique from The Netherlands are using reclaimed 'post-industrial and post-consumer ceramics' waste to create items like lamps and other high-quality products. Their joint endeavor is called Reprint Ceramics and is supported by the EU WORTH Partnership Project. While the former company uses a special 3D clay printing technique, the latter has developed its own process for upcycling household ceramic waste.

Unlike plastic waste, ceramic recycling for 3D printing purposes is an intensive and complex industrial process. But, as with plastic and metal waste, such recycling initiatives reduce the use of raw materials, processing and transport costs, and can repeatedly reclaim broken or discarded ceramic waste. Some ceramics are used to replace metal parts, especially in high-heat-resistant roles. This would help prevent and/or decrease the mining and transport of metal and other Rare-Earth metals. The circular economy of the 3D printing clay industry has the potential to not only enhance household wares and living standards, but also reduce climate change emissions.

Recycling Resin

Even 3D printing resin which uses a liquid photopolymer that solidifies, can be recycled. During research testing, Guoqiang Zhu and his team at one of China's National Engineering Research Centers 3D printed a test part using a castor oil-based photopolymer which was then melted down and reused to print another part. In their post-test quality examination, the part's properties namely ". . . color, viscosity, volumetric shrinkage, tensile strength, Tg [glass transition temperature], polymerization rate, and penetration depth of the recycled resins were almost the same as the original resin," according to Zhu.

What is important here is not all resins can be recycled and not all are developed from renewable sources. This could create an environmental quandary for some 3D printer users and deter the increased use of resin 3D printing. However, the above study used digital light processing (DLP), a widely used product prototyping application. Zhu's team wants to create a "Reversible solid-liquid transition of the printed materials," which is a vital prerequisite for true recyclable printing. Their recycling process of the new castor oil resin starts with the material being completely melted at 90^0 C for 4 hours or 100^0 C for two hours. No catalysts, solvents, or added virgin resin are required during the process or in printing new objects.

Another research team at the Digital Manufacturing and Design Centre of the Singapore University of Technology and Design has also developed a technique that allows for the recycling of photopolymer 3D prints. Like the Chinese team, they also used the liquid-to-solid thermoset polymer as opposed to the typical thermoplastics such as ABS, PLA, PETG filaments.

Their process involved the hardened printed polymer being sealed in a chamber with ethylene glycol and then heated. After some time, the print was melted back into the same original liquid material.

While these are breakthroughs for resin printing, as stated above, this recycling process is not applicable to every resin 3D printing method. Re-printable resin won't be available in the near future unless there is a real commitment to commercialization of the recycled products. However, the technique has the potential to reclaim vast amounts of resin-derived items from printed molds, prototypes, support structures, and models. This enhances the use of DLP printing materials and also creates another avenue for environmentally friendly 3D printing products.

Recycling Glass

3D printing with glass is a relatively recent process developed in 2015 by two organizations that demonstrated glass extrusion techniques. The first was the Israel-based Micron3DP company followed by the Mediated Matter Group at MIT, which they called the G3DP (Glass 3D Printing) project. The processes are highly technical and demanding on the materials they can produce, so for now, glass 3D printing will remain the property of larger commercial and academic entities. However, 3D printing glass with FDM standards is achievable with the latter's G3DP2 process. The MIT team created glass structures that were exhibited at the Smithsonian Design Museum. They also discussed plans that involved glass being used in architectural applications from transparent fixtures to building-size glass structures.

Then in 2017, the Karlsruhe Institute of Technology (KIT) produced a new glass printing method using suspended fused silica in a castable polymer resin extruded by a desktop SLA printer to shape the resin. Further, in 2023, Dr. Jens Bauer's KIT team used a sinter-free process to create nanometer-sized 3D printed glass structures.

Now companies such as Oxman with its G3DP2 process and Nobula with their Direct Glass Laser Deposition (DGLD™) technology are leading a new wave of 3D printing innovations. In fact, while glass 3D printing will accelerate and create new 3D printing commercial opportunities in both the micro and macro sectors from lab-on-a-chip technologies, micro-optics, and photonics, up to aerospace, art, architecture, jewelry, and other industries, Nobula has grander ambitions. Nobula has plans to put a "glass foundry on your desk", a very forward-thinking idea. As of this writing, they have patents pending for 3D glass printing for desktops and glass filaments and they provide glass printing as a service (PaaS). [author note: not to be confused with Platform as a Service, perhaps this should be 3DPaaS].

And since the basis for the 3D printing glass industry will be silica, this leads to the question of glass recyclability in 3D printing. It will be assumed that the organizations above used pristine silica for their experiments and processes. But it would be interesting to know how recycled glass would react to their technologies and if a future desktop glass 3D printer could work with recycled glass. Most likely, the glass would be processed as a service by providers such as Nobula or other recycling filament companies due to the high temperatures required. It may be that only plastic may be a viable recycling option for home users with metal, ceramic, glass, and resin remaining in the purview of higher commercial organizations. Be that as it may, recycling glass for 3D printing still has viability.

Researchers at the Nanyang Technological University, Singapore (NTU Singapore) have created just such a method, which can lead to potential advances in manufacturing environmentally sustainable products and structures from the silicon dioxide, or silica, (a major component of sand) within the recycled glass. In a proof-of-concept test, their method involved combining a specially formulated concrete mix consisting of trade cement products, crushed recycled glass with medium, fine, and superfine sizes, a reduced water content less than normal concrete mixtures, and other additives from which they 3D printed a concrete bench. Further lab tests of the recycled material included compression tests, filament quality, and strength tests. It passed industrial (structural) buildability and (fluidity) extrudability benchmarks. While specific details of the technology and admixture are properties of the university, the potential for other companies to create their own recycled glass techniques would quickly lead to advancements in the process.

The NTU Singapore test has been a revelation in the field. With glass recycling lagging behind other materials in some countries, there is now a chance for glass to be repurposed in the circular/recycling economy with environmental and cultural benefits. I discuss other ideas for glass recycling in Chapter Twelve.

Biodegradable and Biomass Recycling

Biodegradables

Since 3D printing is an inclusive cultural endeavor, there are companies and customers who would be reluctant to use 3D printers due to the real and perceived high amount of plastic used for filament and its wastage and non-degradability. While PLA filament is a biodegradable material, it is still based on plastic and can take years to totally degrade. What other materials can be used that could ease the conscience of plastic-wary would-be 3D printers?

Based in Belgium, B4Plastics bill themselves as a Polymer Architecture technology company with radical design and R&D projects to create ecological and strong materials. Their catalog suite of BioBased Building Blocks (B4) enables safe and pollution-free applications for marine environments, including strong, durable, and biodegradable fishing gear and textiles, which can be recycled, but also mineralize into biomass and biogas. Their process creates new building blocks derived from biomass utilizing "enzymatically degraded plastic waste or paper and cardboard waste" which confers their polymers with "unique characteristics of recyclability, ecology, and unlimited recycling, surpassing fossil carbon at each recycling stage." And for transparency and qualitative purposes, they test their polymers in different environments for fine-tuning and also allow their polymers to be tested by their stakeholders, relevant user groups, and global consumer goods producers.

One of their trademarked products, developed with Trideus, is called Compost3D®, a 3D printing filament. It is composed primarily of natural raw materials which are 100% compostable even for your garden. And in a truly bold statement, they claim that Compost3D® is:

> "the first plastic product ever that you can predict and track in its composting speed on your smartphone."

A filament that comes with its own smartphone app is brilliant. It creates a feeling of trust in standards and product marketability. This is the type of direction needed for 3D printing filaments which is discussed further in Chapter Seven. For now, the smartphone app that B4Plastics has created for its filament is an idea that could be expanded to cover other recycling schemes and incorporated as an industry standard, not just for 3D printing, but also in other industries. Who wouldn't want to know how their building materials and manufactured products fare in the composting and recycling process? It would create an eco-conscious customer-client partnership and inspire trust and the growth of further climate-friendly industries and products.

Other companies are also creating green filaments like Greenfill3D, located in Poland. With a goal to create a zero-waste policy in 3D printing, Greenfill3D boasts a material portfolio of eco-filaments that are biodegradable, compostable, and recyclable. They eschew the use of "forever plastics" (petroleum-based ingredients) in their products which include everyday decorative and home decor items. Their bioplastics are made from blends of biodegradable biopolymers, doped with natural raw materials (like wheat bran, wood, algae, flax, or hemp. With the added use of recycled materials (dubbed rPLA and rPETG, etc), they propose that they are achieving "the least invasive method of modern manufacturing." While they do not sell their filament, they employ over seventy 3D printers to manufacture their specialized serial production.

Lastly, from Massachusetts, 3D Printlife produces and sells filament derived from 'nuisance algae'. Their sustainable ALGA filament is a formulated PLA made from algae compound and an optimized PLA. The algae, once defined as invasive or a nuisance, is sourced from lakes, ponds, and waterways using dissolved air flotation (DAF). This process removes invasive algae from the environment, re-introduces oxygen into the marine system, and the ALGA also releases nitrogen into the soil aiding in nitrogen fixation. In quality, ALGA conforms to ASTM D6400 (a compostable products test) for biodegradability. And for the environment, a proportion of every ALGA spool revenue will be donated to plant a tree in key locations throughout the United States.

Biomass

There are many more companies creating or selling more eco-friendly 3D printing materials. Biomass materials are also becoming more important. The buzzword in biomass printing is lignocellulosic—biomass extracts containing hemicellulose and lignin. In 2022, Eylul Gokce Bahcegul and his team from the Middle East Technical University (METU) were able to 3D print using an alkaline soluble portion of corn cobs in its crude form. No purification, modification, or extra polymers or additives were required. Following the alkaline soluble phase, which forms a thermoreversible cold-setting gel when its water content is partially evaporated (ideally with 17% loss), the gel flows at mild temperatures (45°C) during printing, retaining the intended shape at room temperature on the build platform. After the 3D printing process, the models are immersed in an ethanol bath to permanently fix their shape. The advantage of using cellulose biopolymer is its abundance, biodegradability and renewability. Further uses are being sought for lignocellulosic, hemicellulose, and lignin in 3D printing.

From the US Department of Energy's Oak Ridge National Laboratory, Samarthya Bhagia is developing advanced renewable materials for 3D printing with plant geneticists tailoring properties of biomass, from the plant genes to the final printed products. Bhagia is also looking at lignocellulosic biomass, such as grasses and woody plants with a view to also help reduce carbon dioxide emissions, reduce petroleum-based plastic usage, and support a circular economy. His approach involves developing 3D-printable, biomass-filled thermoplastics with ingredients such as polylactic acid (PLA) which is a thermoplastic made from lactic acid that can be made by fermenting the sugars in biomass. Polylactic acid also has lower net-CO_2 emissions than conventional thermoplastics. By adding biomass from wood or grasses this increases the rigidity of 3D printing composites, reduces the amount of thermoplastics in the composites, and lastly, reduces the weight of the composites. Bhagia's method creates an industry for using biomass wastes from pulp and paper, fruit peels and other food and agricultural crop waste, and small-diameter unsuitable-for-furniture wood pieces. The results will produce low-cost, thermoplastic products possessing the qualities of conventional thermoplastics, but which are also biodegradable and greener composites.

A published report *3D printing of biomass-derived composites: application and characterization approaches* (June 2020) from researchers from the State University of New York College of Environmental Science and Forestry, the University of Tennessee Oak Ridge National Laboratory, and the University of Tennessee Institute of Agriculture states that "There have been 5,100 patents for 3D printing with cellulose since 2015." 3D printing with biomass applications is rapidly gaining ground as researchers and companies seek composites made from biomass, especially lignocellulosic biomass (comprised of cellulose, hemicellulose, lignin, proteins, and other extractives) additionally used in creating biofuels.

One application that the research team are looking into is direct-ink writing (DIW). This type of 3D printing used with cellulose for its fluidic properties relies on the material dispersing well within water to act as a suspension. From this process, hydrogels can be created producing a variety of bioprinting opportunities. Previous research investigated CNF (Cellulose I nanofibers) hydrogels for use in neural bioprinting. They believe that further development of such biomass polymers can result in "smart materials, able to respond, and deform to changes in environment like temperature or moisture levels." Couple those materials with 4D and 5D printing techniques and material science will enter a new age. The researchers also want to further exploit the under-utilized terrestrial biopolymer lignin, second in abundance to cellulose. Lignin has hydrophobic, antimicrobial, and antioxidant properties and can act as an excellent reinforcing agent in 3D printing composites from flame retardant products, anti-aging products, UV ray absorption, and in drug delivery systems.

Biodegradable and biomass materials could eventually out-pace plastics, metals, ceramics, resin, and glass as a major 3D printing material. While there could be concerns overusing cash crops for materials, the upside to using these materials would be their ability to be made from waste materials. Waste not, want not, will take on a new meaning when scraps go from your plate to being 3D printed as a plate.

Plastic Fuels

Recycling should also be about recycling for nature. There are already companies creating artificial trees to sequester CO2, building coral and other underwater habitats from our waste (or surplus discarded resources) for nature to thrive.

Let's be circular in our 3D printing industry. We can both make and recycle the world.

Arguably, the 3D printing industry is part of the fossil fuel industry utilizing tons of plastic for products. At some point, the recycling of plastic filament, unwanted parts, and domestic plastic waste will reach a finite stage of usability and sustainability. So what can we do?

Why not turn the plastic back into fuel? The plastic fuel process is where plastic is converted back into fuel via pyrolysis, which requires heating the plastics at a high temperature and pressure. The water breaks down the plastic and converts it into oil. This would be 'greener' than fresh fuel pumped and refined from the ground. It's not ideal, but the resource is already there to be converted and reduce fossil fuel use.

This innovative technique was used by Washington State University researchers to convert Polyethylene (a component of PETG) by using a catalytic process to create ingredients for jet fuel within an hour. Other products and lubricants can be produced by this method creating a market for cost-effective plastic reuse. By not using the common melting and remolding technique which lowers the quality of the material, this new process can maintain an economic recycling cost and resale value for the recycled plastic and fuels. And as with most proto-designs, the next step will be to recycle other types of plastics. Thus recycling 3D printing products and waste materials would be a sustainable business strategy.

3D Printing Recycling Services

Despite the importance of the recycling and environmental aspects of 3D printing described above, we find there are no widespread practical local services to recycle any of the above more common materials such as plastic, metals, and ceramics. Also, the cost of the equipment to do so ourselves is somewhat out of our price range or the recycling process is an expertise that exists only for large companies and researchers. This furthers my argument for the culture of businesses to change and adapt or reform their print-sell mindset to a cultural-friendly business, which incorporates sustainability. Such actions would make 3D printing more attractive for investment and interest from non-3D printing parties.

But if the 3D printing industry really wanted recycling services, the process would most likely be a subscription and incentive-based service. In this model, subscribers would send in waste materials to drop points or arrange collections for drivers to collect the waste material. The recycled material would then be processed with each waste material identified with their subscriber. Then, for finished sellable products made from the recycled material sent in, the recycler would pay the identified material subscriber a percentage fee or sell the recycled material back to them as filament, powder, or pellets, etc, at a reduced cost for their own production. There would be variations to this, but it would have to benefit both parties for it to work effectively as a proper green recycling industry.

With all the experience gained from the recycling process, we could then feedback data to recycling experiments to create charts of stable recycling processes and materials. But there could be other ways recycling for 3D printing can help communities and other local networks. In the UK, most of the domestic recycling material is actually incinerated for energy production rather than recycled. Such action makes a mockery of recycling programs and would be a disincentive for people to recycle.

A solution could be for 3D printing companies to partner with local authorities in collecting recyclable material to be repurposed for 3D printing. Trials would determine which types of materials would be best suited for recycling. This initiative could be through financial incentives paid for or subsidized on a national or local government level or by the private sector. Residents using the scheme would be sent pamphlets and/or emails regarding the 3D printing services and could be rewarded with 3D printed recycling receptacles made from recycled plastic or metal. There would be public recycling bins for plastic, metal, glass, and ceramic.

Incentives for residents to recycle would include:

- Receiving the aforementioned 3D printed recycling receptacles.
- Gaining vouchers to receive free 3D printed objects whether everyday items, niche products or even 3D printers.
- Winning prizes related to 3D printing.
- Receiving free or low-cost 3D printing lessons.

Further rewards would include:

- Local authorities receiving greater access to 3D printing facilities and services.
- Schools becoming part of the 3D printing community via STEM lessons and recycling programs.
- Local businesses signing up for 3D printing recycling services and sponsoring 3D printing initiatives.

The list would be endless, but we need the demand and incentive to deal with recycling more effectively. With the amount of waste that is lost through non-recycling and incineration, 3D printing can become another outlet for encouraging and fully utilizing recycling in the community while guaranteeing a cultural foothold into households and businesses.

3D Printing Recycling Standards Organizations

Above were some ideas for recycling, but the main issue, as with most things in the 3D printing industry, is that there isn't much industry-standard data on 3D printers and printing materials. Several companies have their own research, technical data, and sales data on recycled filaments and other printing materials. What is needed is the creation of a cohesive recycling system for all materials and for these companies to share and input their data into the system. A codified system would incorporate set criteria, standards, and regulations for processing

3D printer waste, for testing the new material's quality, the long-lasting nature of the material, and its impact on the environment. The recycled materials would be tested, classified, and endorsed by whichever standards authority is established.

Self-regulation of 3D Printing Recycling

As with most industry standards, perhaps it would be best to start small and let companies come together to create and self-regulate their own industry-wide 3D printing recycling system. That way, the companies involved could develop and control their own destinies. This would include codes of conduct, monitoring of legal and safety standards, and maintaining quality assurance compliance for each company and those that subsequently joined. Since 3D printing is still in its insular phase, this may satisfy the industry wary of 'outside interference' from governments or other third-parties. They would be invested in their own success.

However, this system should include both large and small companies, perhaps with international partners, sharing their recycling materials knowledge. While each company in the self-regulation system should be nominally equal, the avoidance of conflicts of interest, internal cabal formations, and withholding of data should also be monitored by an external agency. Transparency of the self-regulated system would ensure advancements in 3D printing recycling would benefit more people outside of the 3D printing industry. With the development and evolution of self-regulated 3D printing recycling processes, these may morph into national or international groups.

National 3D Printing Recycling Agencies

Perhaps national organizations in the US or the UK could develop standards for their printers. There's the American National Standards Institute (ANSI) and National Institute of Standards and Technology (NIST) in the former and the

British Standards Institute (BSI) in the UK. I conducted quick searches through each of the sites to discover how each was handling 3D printing standards, especially in relation to recycled materials for 3D printing.

ANSI has over twenty-thousand entries relating to the terms 3D printing and additive manufacturing. In collaboration with America Makes they have published the Standardization Roadmap for Additive Manufacturing, Version 3.0. Their goal is to define "the current and desired future standardization landscape for additive manufacturing (AM)." Their focus is on the industrial 3D printing market sector and they have identified 141 standardization gaps (including 60 new gaps) that they desire to be adopted throughout the sector, including: design, precursor materials, process control, post-processing, finished material properties, qualification and certification, nondestructive evaluation, maintenance and repair, and data. While recycling is not specifically listed, there is no doubt that the future development of 3D printing standards will have to incorporate provisions for the recycling of materials for the use of 3D printing materials.

NIST has around 7000 entries with the searched terms 3D printing and additive manufacturing. Adding them with recycling does not yield the expected results though a few entries on publications were investigating the properties and testing standards of 3D printing materials.

Unless I was using their search facility wrong, the BSI only seemed to have less than 20 entries for 3D printing and additive manufacturing. That was rather disappointing, and I hope the UK has a more robust standards organization that will facilitate the requirement to use 3D printing to its fullest potential, especially in conjunction with recycling waste materials for future 3D printing use.

Of course, new 3D printing-specific recycling organizations may be created, filling the gaps established and/or traditional organizations could not bridge.

But, having such organizations to learn from or branch from would be an ideal start.

International 3D Printing Recycling Organizations

For a global solution, the International Organization for Standardization (ISO), which develops and publishes International Standards could be approached for the role. The ISO has developed over 24, 598 International Standards which are cataloged for reference. There are around 130 3D printing-related submissions. However, the main ISO standard is ISO/TC 261 Additive Manufacturing (created in 2011). It is part of the Technical Committees and standardizes aspects of 3D printing processes, terms and definitions, process chains (hardware and software), test procedures, quality parameters, supply agreements, and other fundamentals. The standard also publishes standards, with 27 completed, and another 32 under development. From what I observed, none were specifically related to recycling materials for 3D printing.

Interestingly, there was one article pertaining to 3D printing and using new-generation plastics where "innovations in recycling technologies will make manufacturing the materials of the future more sustainable." Having ISO credentials or a similar international seal of approval would set up 3D printing waste recycling as a credible and viable industry. It could also provide vital non-negotiable terms to achieve these standards in any 3D printing recycling contract and integrate recycling into future 3D printing industry and environmental management systems.

At the moment, what we're seeing is a lot of commercial probing and innovative academic research. What we need are the metal, ceramic, resin, and glass 3D printing companies to pick up on these programs, invest in more recycling initiatives and achieve these lofty ambitions while driving the industry forward. We need the business infrastructure, research, testing, and commercial nous to bring these recycling advances to the general consumer. While the 3D printing industry can establish its own recycling standards at some point an international standard will be required and the ISO would be best suited to maintain and enhance the work. Thus, from disjointed recycling services, we can create an overarching system leading to qualitative and quantitative cultural standards within the 3D printing industry.

CHAPTER SEVEN

3D Printing Media

3D printing is at a cultural disadvantage within the media. Despite there being millions of 3D printers around the world, one of 3D printing's limitations in further cultural dissemination is that it is mainly advertised and sold online. 3D printing is seldom seen in 'mainstream' news or on TV, unless as a sci-fi-related tool. There are magazines, YouTube shows, and websites devoted to 3D printing, but these are for niche audiences and not for casual customers. So how can 3D printing be made more appealing and visible to the general public?

You can YouTube and document the advances and benefits of 3D printing all you want, but the general public wants to know how it's beneficial to them now. Unless the product is in front of people ready to solve their problems on a daily basis at the push of a button or verbal command without the need to take a course, then 3D printing won't become an indispensable appliance in the household.

Below, we take a look at 3D Printing education, TV advertising, online shows, magazines, apps, AI, and social media and how they can influence the future of 3D printing's journey into the public limelight and your home.

3D Printing Education

As a cultural rite of passage, education is a valuable tool for propagating and preserving ideas for the future. In the course of trying to set up my company, I developed an online course to add to the growing corpus of 3D printing courses. Below, I present an outline of my course and workbook ideas in full, so you have a visual flavor of what I was trying to achieve when I tried to pivot my company toward courses. Was I on the right track? What could have been improved? Of course, in the spirit of networking, there's also the opportunity for you to riff off the materials and create your own courses and workbooks. But soon, 3D printing may become part of the school curriculum.

Encouragingly, China has established the world's first 3D Printing College, the Baiyun-Winbo 3D Printing Technology College in Guangzhou. The government also plans to equip over 400,000 schools with 3D printers enabling 3D printing skills from an early age. Such students, even rural ones, can then take their skills out of the classroom to their homes and later working lives, whether as individual entrepreneurs or at companies. This national-scale learning path and the decentralized online courses are blueprints for starting 3D printing careers, securing much-needed tech skills, and for future-proofing manufacturing jobs and services. Other nations would do well to follow suit.

One such recent 3D printing educational pioneer is the Dubai Municipality, which has launched the world's first certification program for the 3D Printing construction sector. Their goal is to standardize and streamline 3D printing procedures within the construction industry, creating a standard for the concrete mixes used in the city. The Dubai Central Laboratory with its experience in evaluation, conformity auditing, and lab material testing will ensure quality control in the standard of the certification. This is the sort of action required in other industries using 3D printing. It's not enough to create your own standards, but to make sure it can be used to create an industry-wide standard or a suite of standards so cities and countries can share and update their techniques for their purposes.

Of course, the immediate future of 3D printing depends upon people learning how to use the technology. My aim is for this step to be unnecessary in the future so that 3D printing is just another household appliance like a microwave, photocopier, or gaming suite. Did you have to take a course to learn how to use them? No. And in the future, it will be the same for 3D printers. We just need the right user interface, intuitive applications, and incentives to make 3D printing available for all.

But for now, we have to take courses or learn intuitively, whether in a classroom, at work, or self-taught at home. After creating my company, I found I had to pivot its focus and include a fund-making element that also served my current and potential customers. I had to future-proof myself, and the way to do that was by creating a beginner's course for 3D printing. And more than that, it had to be online.

At the time, the world had been gripped by the Coronavirus pandemic and so industries had quickly switched to online services to carry on business. Plus, with the online option, I could reach customers around the world. After researching different courses and teaching techniques, I created a course that would have been a mixture of theory and practical covering the basics, with the option to add more courses. But for various reasons, explained in Chapter Thirteen, the course was not delivered. However, below is the version of the course I created.

3D Printing Beginner's Course Outline

Description

We will be starting courses on 3D printing, 3D modeling, software use, and more. This is the first course designed for beginners in 3D printing, its working processes, advantages, and how you can create your own works for pleasure or business.

This course also gives you the opportunity to buy your own 3D printer for the course, or as an extra printer if you already have one, or as a gift for others, or for your business. It is a great investment to further learn the ways of 3D printing for yourself. We're here to take the stress out of learning 3D printing with easy-to-follow courses. Let's dive in!

THE FUTURE OF THE 3D PRINTING CULTURE

Syllabus and Layout

Online course, video and Powerpoint slide show, theory and history of 3D printing, practical use of 3D printer, and software usage.

We'll teach you 3D printing so you can utilize the best way to apply for design roles or set-up a creative business. Or even if you're just looking for upskilling or key technical skills to make your CV and portfolio stand out from the crowd.

Our 3D printing course will teach you the ins and outs of this exciting new technology. By the end of the course, you will be confident in using a 3D printer to create your own models. You will be taught how to design models using 3D software. Your course purchase includes:

- 1 hour on-demand video, beginning with theory presentation.
- Learn anytime, anywhere at your own pace.
- Full lifetime access.
- Access on mobile and laptop.
- Certificate of completion.
- You will only be charged once for the purchase of this course.
- Any future updates to this course will be automatically distributed to you at no additional cost.
- 30-day money-back guarantee

Course content

4 sections • 4 lectures • 1-hour total length

What you'll learn:

Module 1. Introduction - The theory of 3D Printing – 1 video

- Overview - This is an overview of what the course will be covering.
- Theory and background of 3D printing.
- Types of consumer-level 3D printers.
- What is FDM printing?
- Limitations of printing.
- 3D Printing materials.
- Exercise - Understanding theory confirmation.

Module 2. Practical 3D Modelling – 1 video

- Introduction to Sketchup (or other design software).
- Basic modeling principles.
- Considering print limitations when designing.
- Modeling to scale.
- Layer wall height.
- Infill, Shell, Supports (outline, skirts, rafts, etc).
- Groups and layers.
- Best print orientation.
- Downloading models.
- Slicing your model - Master proper slicing software settings for best the print results.
- Basic Cura Controls.
- Preparing STL files and sending them to the 3D printer.
- Understanding G-codes.
- Troubleshooting.
- Exercise - Design your own models to be printed.

Module 3. Practical 3D Printing – 1 video

- Parts of a 3D printer.
- Setting up a 3D printer.
- Effectively leveling the print bed.
- Setting bed and nozzle temperatures.
- Set up their printer for proper first layer bed adhesion.
- Pressing Print.
- Monitoring your print.
- Troubleshooting.
- The Finished Print.
- Exercise - Complete final print for course completion.

Module 4. Exercises – 1 x video

- We will show you how to access libraries of thousands of free ready to download and print designs.
- Design and print your own decorations, jewelry, and practical household items.
- Final tips and advice.

Upon course completion:

- Certificate of completion for your CV and LinkedIn files.

Extras

- Opportunity to purchase your own 3D printer.
- Opportunity to purchase 3D printing workbooks and report books.
- Access to private Facebook course group.
- Free digital course reference PDF for revision and development.

There are many more options that could have been included with extended module numbers and time sessions, plus intermediate and advanced classes, all at affordable prices, but this was the core beginner's course.

Workbooks

Project/Journal/Organizer Books

As with any course you'd need a workbook to record your progress and print works. I came up with a couple of designs, where you could log 25 or 50 projects in a physical book, though no doubt there are digital formats that could be used. The first type was a 3D Printing Project/Journal/Organiser for 25 projects, followed by a 50 projects version. The index would contain a guide for your project numbers and model/file names. And as with the course, these can be used as templates for your own workbooks.

The books were formed of 4 pages repeated until the book was either a 25 or 50-project A4-sized book:

- Prep page – the upper half of the page was the Print Prep form to record all your settings before printing so results could be measured and repeated or changed, accordingly. Under this was a lined Prep Notes area for more details.
- Isometric Grid page for drafting models.
- Print Report page on the upper page to document how the print or project turned out and other comments. Under this was the journal/ organizer page so you could chart your projects, prints, or any other 3D printer-related business on a daily, weekly, or monthly basis.
- Supplementary Notes, Ideas, & Drawings page; a blank page for more creative endeavors.

The four pages are illustrated, below:

Print Prep

Model /file name: _____

Date: _____

Printer:
Brand: _____   Printer: _____

Filament:
Brand: _____   Material: _____   Colour:_____

Nozzle temp:_____   **Bed temp**:_____

Layer Height: _____   **Shell thickness**:_____

Raft: Y/N _____   **Supports: Y/N** _____   **Infill %**_____

Speed:
Print:_____   **Travel:**_____   **Infill**:_____

Print time:_____   **Filament amount used**:_____ g/Kg

Prep Notes:

_____
_____
_____
_____
_____
_____
_____
_____
_____
_____

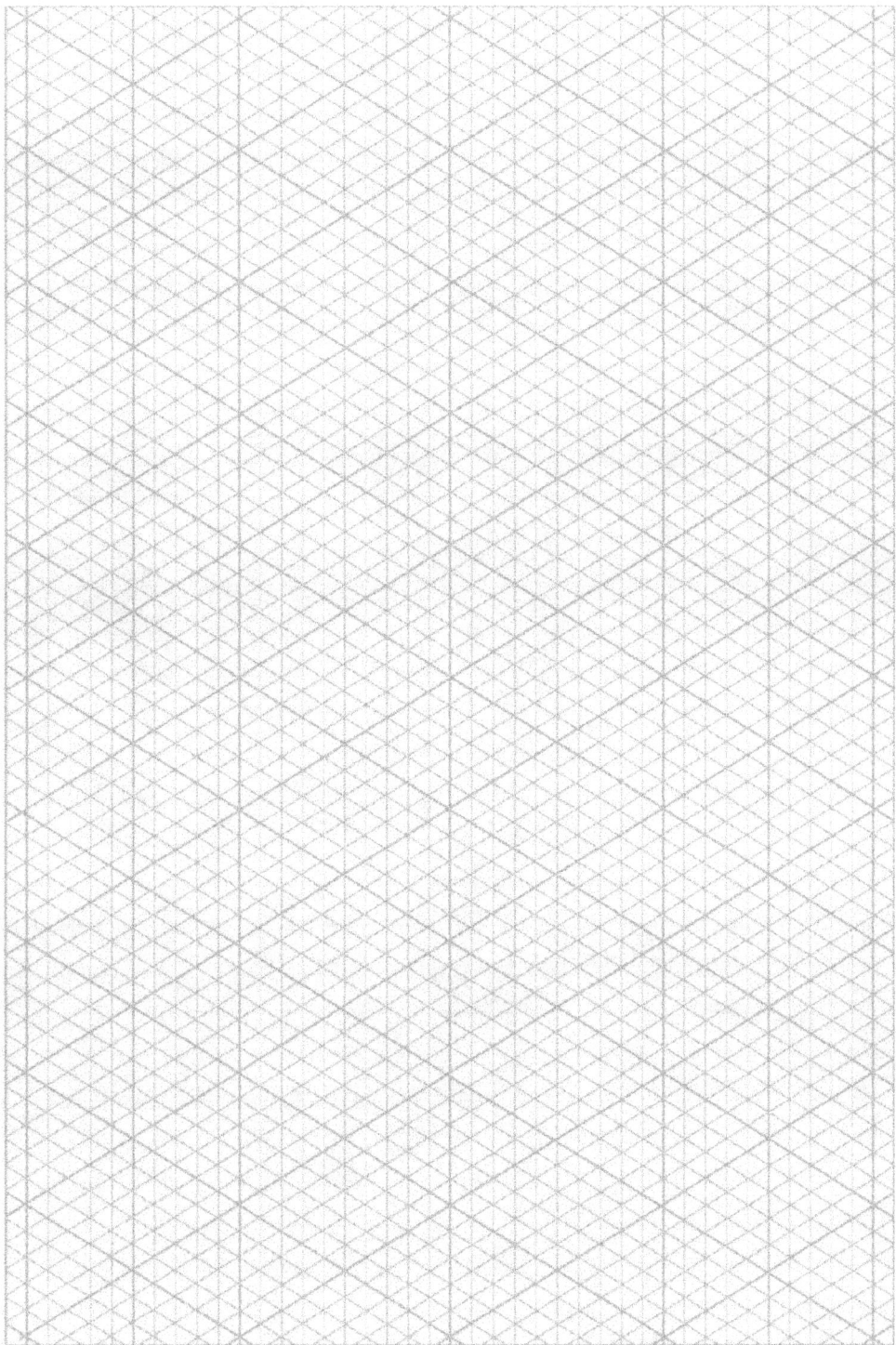

Print Report

Comments (successes/challenges/errors/lessons learned)

_____
_____
_____
_____
_____
_____
_____
_____
_____
_____
_____
_____

Organiser/Planner

Month:_____

Mon	Tues	Weds	Thurs	Fri	Sat	Sun

Supplementary Notes, Ideas, and Drawings

Report Books

My other workbook was called a 3D Printing Report Book, which had the same split-page Print Prep and Prep Notes along with a split-page lined Print Report and on the lower half, a blank Creativity Space. This allowed for more creativity on the projects leaving out drawings and journals, for expedience.

The workbooks would be an extra revenue stream and give the students a suitable and course-enhancing physical bonus. The courses would be advertised online, in magazines, and I wouldn't have ruled out door-to-door leafleting, with 3D printing on the junk mail junket.

Never mind the fact that 3D printing is being taught in schools and places of higher learning, with the increase in E-Learning and the advent of Edu-tech, 3D printing will be able to be introduced to people around the world whether they have actual hands-on experience or not. Virtual 3D printers can be set up whether over a Zoom lesson or on the burgeoning metaverse. Once you have touchscreen availability you will be able to use a 3D printer remotely and take courses or rent space in an academic venue with 3D printers or a 3D printing farm for your homework or for practical experience. Out of all the new exponential technologies, 3D printing will be one of the most decentralized and it's still in its nascent stage. So, how can it be effectively advertised?

TV Advertising

You see every other technology like flat-screen TVs, mobile phones, computers, games consoles, and more sold on every media outlet, especially TV. Why not 3D printers and their accessories? A 3D printer seems to be a mystery to the public and it will stay that way until there is a determined effort to publicize its existence, teach the technology, advertise its benefits, and sell the products on a visually-based media, like TV.

Additionally, when was the last time you saw an ad on TV that stated the product had been 3D printed or was partially 3D printed? Never! That includes vehicles, sports and medical equipment, furniture, and other tech products. None are shown to have actual or potential 3D-printed parts. You watch a lot of programs about being environmentally conscious, but none show how 3D printing with recycled plastic waste (like old fishing nets and discarded debris) scavenged from the seas can save marine life by limiting their ingestion of plastic and help build new artificial or support damaged marine habitats like coral. Even if 3D printers and equipment aren't sold directly, there is still room to reference 3D printing where used.

Some advertisers and entrepreneurs may say TV advertising is old-fashioned, overrated, and not cost-effective. But place a 3D printing ad during the Super Bowl, the Olympics, or even in cinemas before films and you've reached a whole new demographic worldwide. It's worth taking that high financial risk advertising on a medium a lot of people still view. Even previously ad-free streaming sites are now featuring ads, so target them as well. Get 3D printers in front of peoples' eyes and you open up fresh and growing markets.

I bought my first printer from Amazon because it was the same type I was trained on, so I was familiar with it. If I hadn't had that introduction, I would have looked online for comparison sites and then gone straight to Amazon. I wouldn't have gone to a brand company as I would have with other tech products. Currently, I see a lot of 3D printing newbies on Facebook asking which 3D printer they should buy, because they have no other visible outlets to learn from or trust, so they turn to the 3D printing community. You would not get this with other tech products as they're constantly advertised on TV and you know the brand and attributes with a high trust factor in the product.

3D printing has to tap into the home advertising market to drive publicity and sales otherwise 3D printing stays a novelty machine for the domestic realm and an expensive corporate showpiece. Some may feel the 3D printer is not a fully

-developed or realized technology for widespread household usage yet. But, I say this is because the 3D printer is still in a niche phase so its development is stalled until new markets and uses are opened, hence the cultural domain I am creating. This would mirror initial mobile phone and personal computer design and market growth once more of the population bought into those products. Advertising drives not just sales, but also design and user-friendly features. So, start advertising 3D printers or products with 3D printed elements on TV.

How? Let's look into examples where major food outlets such as McDonald's and Coca-Cola, and a global sports championship, Formula One (F1) could do more for marketing 3D printing in their products on TV.

Coca-Cola and McDonald's are among the growing number of global corporations that use 3D printing in some of their marketing plans. However, their strategy still seems like a gimmick. While 3D printing gives them far more advantages over traditionally manufactured products, increased design options, increased speed of production, and the capability to print large orders locally, their efforts have hardly cut into the minds of their consumers. In other words, there's a cultural disconnect between what these companies are trying to achieve for their consumers (e.g. added value and maximizing product experiences) and making their consumers aware of the benefits of 3D printing. This is where their visual marketing should be in full effect to enhance the tactile effect of the product. For instance, you see the advertised 3D printed product on TV or an online ad and want to experience its physical nature, whether by eating it, using it, touching it, or owning it.

In 2013, a Coca-Cola subsidiary in Israel, used 3D printing in a competition to advertise its mini-sized bottles and increase awareness for their campaign and brand loyalty. Using a mobile app, customers could design a digital 3D model of themselves. Selected winners were then invited to the Coca-Cola factory to be scanned and later presented with their 3D-printed mini version of themselves. While customer engagement and interactivity may have increased, and while you can still buy mini coke bottles, the 3D printing element may be lost in the shuffle. It's a fun, but brief experience. For actual tangible Coca-Cola products, why not 3D print their bottles, cans, or crates, and other physical products they produce to enhance their recyclability and biodegradability reputation? The 3D printing material could have already been recycled or be a biomass material. You'll then be putting 3D printed products, possibly creating a collectible item, directly into customers' hands linking a cultural icon with 3D printing.

An example of McDonald's use of 3D printing in the UK was its 2016 'Always Working' campaign where they created hundreds of 3D printed characters and half a dozen sets for a sixty-second animated stop-motion video advertisement. The 3-month endeavor required a team of fifty people to achieve. McDonald's felt a 3D printed stop-motion film was a better human-touch story-telling medium than using computer-generated imagery (CGI). Did you hear about this video? Neither did I! And even if I had seen the ad, I wouldn't have known the film was created with 3D printers, unless you knew the back-story. Most people wouldn't delve into TV ad creations, so why didn't McDonald's hype up the 3D printing technology? They haven't enhanced the consumer experience or added value through advertising with 3D printing. If they want to connect with their consumers and if production costs permitted, they would be better off 3D printing all their cutlery (though they have shifted from plastic to wood utensils in the UK and Ireland in 2022), plates, cups, trays, and kids' toys and market them as safe, reusable, recyclable PLA. They would further discourage the one-use, throwaway economy. Stick these 3D printed facts on their meal boxes and on their TV ads' small print and people will become accustomed to 3D printed products, because soon enough McDonald's will be 3D printing their meat-free and real meat burgers, and more.

As we move from the food and drink industry, we investigate 3D printing in another great cultural pastime: sports. As a high-profile sporting event that uses 3D printing, even Formula One has a way to go in conventional advertising.

The first team to use 3D printing on their cars was Renault in 1998. Nowadays, F1 teams use 3D printing for prototyping and R&D for modeling aerodynamics, building chassis, hydraulic and engine parts, and other equipment. The efficiency from 3D printing reduces car weight, material costs, and the industry's carbon footprint dove-tailing with F1's long-standing Net Zero Carbon policy. Teams such as Alfa Romeo partner with 3D printing company Additive Industries; McLaren Racing uses Stratasys; while Alpine is with 3D Systems; and Ferrari has 3D printed sensors on its cars' front wing.

Despite this forward-thinking process, Formula One has decided to produce a high-end 3D printed product for its high-end customers, namely their F1 Fragrances Engineered Collection. This consists of a fragrance bottle housed within a web-like exoskeleton outer structure 3D printed with technopolymer resin. When the fragrance is finished, the core perfume container can be removed and replaced, like an ink cartridge in a pen. The designer's vision was to imitate F1's goals of sustainability and efficiency by printing only what is necessary for the outer structure similar to an F1 car. But how many of the purchasing customers will appreciate the role 3D printing played in creating the product? How will the product resonate beyond F1 customers and fans? Will it encourage more companies to use 3D printing for packaging? This is a collectible niche product for a niche audience. For 3D printing to be culturally relevant in the mass advertising of sport it should be represented more openly for more people.

Of the current ten F1 teams, only three seem to have publicly announced 3D print companies as a stated partner or sponsor. Other teams may use 3D printing with in-house teams and or use companies not obviously listed. Moreover, in checking photos of the cars, I could not discern the names of the 3D printing companies on the cars' chassis. But also, who pays attention to the score of sponsor names racing around the track at 200mph? Do you check out the names of sponsors on paddock walls or on the drivers' racing suits? F1 teams and commentators always talk up the chassis carbon-fiber and tire compounds of the cars, but rarely mention 3D printing, which is an advantageous technology. You'd think that with a huge global viewership, the 3D printing companies involved would want their products and efforts publicized and recognized across the global F1 audience much more so than through a fragrance bottle. Just as F1 technology filters through to road cars, traditional car manufacturers should also be able to increase their 3D printing usage. Eventually, we will see advertisements that some production vehicles are entirely 3D printed and/or possess 3D printed parts, especially when those parts come as standard. This will reveal to customers the advantages of 3D printing in their everyday transport, commuting, and leisure lives.

When I brought this subject up on social media, the most reasonable response was that trying to advertise on F1 equipment was incredibly expensive. Well, of course it is, but maybe that points to a deficiency in the 3D printing industry that even the biggest companies are not cash-rich enough to contemplate advertising in such a sporting arena. That points to a bigger problem with 3D printing corporations as a whole, which I take up in Chapter Eight.

I'm sure McDonald's, Coca-Cola, and F1 have reams of literature on their partnerships with 3D printing which would be great for print advertising, but as the saying goes 'a picture is worth a thousand words'. With the examples above, there's no point in using the advertising TV space and time with 3D printing technology while keeping it behind doors. Advertise the s**t out of it. From high-end vendors and products to mid and lower-end ones, flaunt 3D printing if you want to stand out from the traditional manufacturing crowd, even if 3D printing is mentioned in the small print at the bottom of the screen. However, all is not lost with Coca-Cola who are now using 3D printing to create bottle tops from 'skyprinting' as discussed in Chapter Ten.

There are numerous competitors to McDonald's and Coca-Cola who could benefit from 3D printing and advertise that fact on TV. And from F1 to other sports, 3D printing can manufacture athletic footwear, tennis rackets, cricket and baseball bats, field hockey and ice hockey sticks and pucks, shoulder, elbow, and shin pads, helmets, mouth guards, skateboards, javelins, etc, and other equipment, including goals posts for soccer, American football, hockey, and rugby. And the 3D printing company's name would be printed on the sports equipment for everyone to see on global sports competitions year-round on TV. Fans and customers could buy the sporting products on sale, as advertised on TV of course. Your cultural connect to 3D printing.

In fact, just as I was about to go to print with this book, I was watching the TV show *NFL End Zone* hosted by Cori Yarckin. She was interviewing Norman Vossschulte, the Philadelphia Eagles' Director of Fan Experience & Sustainability who was explaining how Lincoln Financial Field stadium was 100% supplied with renewable energy sources, including 40% by solar paneling around the stadium and 60% bought renewable energy, along with other sustainable initiatives they undertook. But he also mentioned that all of their plastic bottle caps are collected and through their partnership with plastics company Braskem, they 3D printed a six-foot tall replica of the Lombardi Trophy, which is exhibited in their main lobby. While not a TV ad it was a great advertisement for and exposure of 3D printing within sporting culture on TV. How many other NFL teams or other sports organizations around the world can boast such green credentials alongside a positive 3D printing experience?

Perhaps it will take a new sport altogether to embrace the technology and benefits of 3D printing. Ahead of the 2023 F1 Austrian Grand Prix, three enterprising individuals took part in a jetpack race. The exterior of their packs was 3D printed using SLS techniques. These are the type of jetpacks you may have seen online where three small engines are housed on the racer's back for thrust with two more on each arm rig (which also hold the hand controls) for maneuverability along with their body motion. They have the potential to be faster than F1 cars, but more than that, they could end up being legal personal transport in the future with a myriad of options for personalized 3D printed exteriors. Now, who wouldn't watch such a futuristic sport and/or use an alternative to the perennial always near-yet-far flying cars? The jetpacks may be at the intersection of sports and transport technologies, but 3D printing can constitute an integral part in the establishment of a new industry; an exciting cultural phenomena in the making.

Of course, this is only about advertising 3D printing on TV. But, in order for it to work, these companies have to make 3D printing meaningful to their customers and not as a marketing gimmick. McDonalds, Coca-Cola, F1, and other businesses could have valid production and economic reasons for not 3D printing more wares and not publicizing their 3D printing efforts, which I can understand and respect. After all, I am just airing ideas, questions, and challenges. It's how inspiration and innovation begins. So, if not TV advertising, then how about whole TV shows about 3D printing?

TV Programs

There have been documentaries like Print the Legend (2014), The Future of 3D Printing (2015) and a plethora of 3D printing shows, documentaries, 'How To's' and other vlogs, ads, and media on YouTube and other online sites, but not on mainstream TV. The technology still constitutes a niche market leading to a dearth of 3D printing-related activity and marketing on TV.

Where is the David Attenborough of the 3D printing world studying the Creality Ender or the UltiMaker in their native habitats? Will a MasterChef contestant become champion by creating a 3D-printed 3-course meal? And will the World Cup be won with a 3D-printed football? The last is probably more likely within a decade.

Most people will see a 3D printer on a sci-fi show or film leading them to think it's still a sci-fi prop. The ultimate futuristic 3D printer is Star Trek's synthesizer/

replicator dispensing food and other items. We're not quite there yet, but it would be a great advertising tool to suggest that 3D printers are the rudimentary ancestors of the replicator. But we need people to publicize that and generate the buzz for learning about, using, and buying 3D printers.

We need new TV shows, documentary style or reality TV shows like MasterChef/Bake Off meets 3D printing or engineering shows. People love competitions and reality shows so why not combine that with 3D printing? We need an 'Everything 3D Printing Channel', where people can go and stream 24/7 for one-stop-shop 3D printing access.

And while I have mentioned documentaries, news, and reality shows, there can also be fictional shows, sci-fi or not, whether set in the present or future with 3D printing an essential part of the show. We can have fictional shows depicting:

- The problems and triumphs of a 3D printing start-up.
- The daily tribulations at a 3D printing farm fighting a 3D printing or traditional manufacturing competitor.
- A soap opera character who routinely uses 3D printing.
- The adventures of a team of 3D printing engineers on the moon.
- Superheroes using 3D printers for their costumes and equipment.
- Conmen using 3D printers to replace stolen treasures and stay one step ahead of the authorities.

The list is endless. We need visibility to drive 3D printing forward.

My Pitched TV Shows

In fact, I did just that. I devised two TV shows for a TV production company. Here is my intro to them:

I would like to propose 2 x TV programmes which combine the world of 3D printing with reality TV and competitive shows with the working titles of: **3D Print Wars** and **Make It For Me**.

3D printing (or additive manufacturing) is a multi-billion-dollar industry, dubbed the new Industrial Revolution with worldwide access and applications. More than ever, 3D printing is gaining publicity and traction with companies offering multiple 3D printing services in mass production, selling, and marketing the products which will remake the world.

And while there have been documentaries and YouTube shows (notably 3D *Print the Future* (2017) and *3D Print Masters* (2019)) regarding 3D printing or featuring 3D printing in science fiction media, the former invites a niche audience and the latter showcases 3D printers as a sci-fi tool. The overall effect still leaves the general population not fully comprehending or appreciating the implications of the technology and the profound effect it will have on society.

3D printers may significantly replace traditional engineering and manufacturing processes and companies. Individuals will be able to innovate and produce their own wares at home or at work with the added incentive of being able to sell their own products.

Hence, the first TV programme *3D Print Wars* will combine the audience's love of engineering and design reality TV with competition to 'show off' real-world 3D printing skills and to make 3D printing 'sexy'. There will be opportunities to learn about 3D printing, how it works, what can be made, and its future applications and consequences for society.

And the second programme *Make It For Me* will attract audiences who desire quirky design and craft programmes with bespoke, custom, and unusual visual and physical creations. But, here, they will be created by a 3D printer, whether furniture, shelters, artwork, clothing, etc.

The competitions will be challenging and fun, with the chance to win prizes from a client or business or showcase their abilities to the public. With the 3D printing industry growing every day, *3D Print Wars* and *Make It For Me* can also branch out with junior, professional, and international spin-offs.

This is engineering at the cutting edge put through its paces by clients eager to have a specific product designed and printed in such industries as medical, food, building, transport, and even space. With cooking, baking, sewing, gardening, carpentry, engineering, and painting skills, etc being illuminated on TV and with the gaming of apprenticeships, singing and dancing talent shows, and bidding for a stake in a company, it's time for 3D printing to join the cavalcade of competitive reality programmes and give the audience a glimpse of the future.

I have since created a third show in July 2023, a variation on the above, provisionally titled *Printer Winner*. My brief notes were:

- Straight 3D printing contest.
- Go large with Monster Printer Winner.
- This is about the personalities, the mavericks, big shots, and entertainers with their industrial 3D printers.
- This is about spreading 3D printing culture to the max.
- 12 hours to design, slice/code, print, and post-process their object.

The above was followed by more detailed work-ups. What's in it for the production companies? It makes their life easier to find new content. However, while the shows solve a problem for society and advertise new technology, which they love, independent production companies loathe taking submissions from external sources lest they decline it, produce a similar show, and get accused of theft. Be that as it may, the full work-ups of the three TV shows are in the Appendix, so feel free to credit me if you use my ideas to make the shows!

There is a drawback to watching 3D printing, however. It's boring. It's great watching the first few times, making sure the base sticks and key elements are printed correctly, but watching several hours of printing can be dull. So, for each of my work-ups, I suggested a time-lapse montage so the key parts of the print can be watched, thereby negating the lengthy viewing process. And while the printing is ongoing the hosts can do voice-over action or interview the contestants.

Ultimately, to help 3D printers become ingrained within culture, we should embed them within popular TV shows like soap operas, police procedural shows, cooking and arts & crafts shows, things that are non-scifi. That way the cultural visibility and impact would be more apparent.

Online 3D Printing Content

Other ideas would be to start a dedicated 3D printing content-filled channel (just like shopping, travel, cooking, and crime channels) whether on terrestrial TV, YouTube, or a streaming site with all the indys placed on the channel with member subscriptions. The channel would include 3D printing NEWS (News, Entertainment, Weather, and Sports) relating to 3D Printing. 3D printing impacts all four of these so a show modeled on the NEWS, fictional shows, and other original programs with indy YouTube presenters filing their own exclusive content would be a massive step forward. An all-inclusive site with 3D printing content already created and with new content to drive maker and printing publicity, and knowledge, would enhance 3D printing's profile exponentially.

What other kind of content would there be besides the news? You could have the usual 'How-To' shows, special build and engineering/design shows (like my own pitched shows), merchandise sales, documentaries, international shows, competitions and exhibitions, and even game shows. The more content, the more ideas and development for the 3D printing industry.

Disadvantages or restrictions online/independent producers may face on a dedicated channel could be structured and scheduled programs with deadlines for airing. While shows can have any content and vary in length there may be advertising tied into a show's ratings. However, this may serve to raise standards and production values for 3D printing shows. There may be objections to subscription fees for content like on Netflix and Amazon when viewers can watch content for free on YouTube. However, the idea is to create a medium through which 3D printing's profile is raised, valued, and given time to mature as a cultural asset. 3D printing would thus gain more exposure and be endowed with personality advancing the knowledge of 3D printing to anyone interested in the technology tuning in.

For views on my TV shows and dedicated 3D printing channels ideas, I looked through LinkedIn and contacted a well-known host of a 3D Printing show on YouTube. I told him I was thinking of ideas to advance 3D printing within the decade which also incorporated advancing 3D printing advertising, and its TV and online presence with a TV/online 3D printing channel bringing together various independent 3D printing content producers in one place. I wouldn't see each content producer in competition with each other like on YouTube trying to get views, but more like a cooking channel or sci-fi channel which has multiple same-genre shows.

I didn't get a response even though he had read the message. I realize I am basically cold-calling strangers for opinions here and elsewhere in this book. Though I had hoped the community of 3D printers would offer some of their learned views on the industry. There's still a ways to go on this journey to creating the lifestyle and future of the 3D printing culture.

3D Print Magazines

There are dozens of 3D print magazines on the market ranging from general information to more specialized niches, whether in print, online, or both. However, magazines don't really make great in-roads in the media in general or eclipse the digital world. And if trying to penetrate non-industry markets to publicize 3D printing then another strategy is required.

Quick-thinking writers do this a lot and that is to break down the larger piece of content and parse it out in smaller digestible chunks for the general public on forums, blogs, social media pages, and sites not necessarily associated with tech. Students, stay-at-home parents, influencers, investors, and craftspersons would be the natural targets to increase publicity for 3D printing. Building communities outside of the industry is key.

3D print magazines may just have a one-time use if a new product is launched or there's a special feature or comparison between several models. They may be kept as souvenirs or consulted for research. But generally, they will have an online equivalent.

The best use of a physical magazine would be as a glossy 'freemium' brochure for your company or products given out at events. They would be a calling card, a way to be noticed and contacted. Freemium marketing materials can be 'sold' free online for the cost of postage and handling. You earn a small revenue from a supposed free product. So, trying to attract a client or customer to your 3D printing company? Offer to send them a 'free' magazine.

3D Printing Apps (3DApp)

At the moment, 3D printing apps seem to be centered around the usual handful of technological suites of services, such as:

- 3D modeling, drawing, design, sculpting, etc.
- 3D printer remote control for decentralized manufacturing services.
- Scanning.
- Model file sharing, trading, and slicing.
- 3D printer and model sales.
- 3D print forums and groups.
- 3D printing news Apps.
- Remaining filament amount and print cost counters.

In the future, I'm expecting to see apps for 3D printing games, courses and learning, 3D printing streaming sites, metaverse features, workbooks and project organizers, and sales of 3D printing furniture and fashion. Many companies will be jumping to start their own apps and be the Amazon, LinkedIn, or Facebook of the 3D printing world.

Remember the B4Plastics mobile app? How about increasing the information on a 3D printing recycling App including:

- Which materials can be recycled from print waste?
- Which non-3D printing materials can be recycled as print material?
- How to recycle 3D print materials whether by yourself or using a service?
- Multiple company information on the quality of their recycled materials.
- The pros and cons of recycling.
- The economics of recycling.
- The carbon footprint of 3D printing recycling.

Let's call it the 'threecycling' app. And of course, that would only be the start of the new 3D printing app revolution.

You will also be able to download a 3D printable gift app. Cards and ecards dominate the greetings industry, but how about an e-card or voucher with a difference? Simply download a 3DApp and activate any participating printer anywhere in the world remotely. You would be allowed to create a design, 3D print, and post/sell your 3D printed gifts or cards anywhere, anytime.

A 3DApp can also be used as an aggregator, listing all 3D printing apps such as Thingiverse, Tinkercad, My Mini Factory, etc to make it easier to print for yourself. Apps will open up new avenues for optimizing your 3D printer's green footprint and providing transparency. And with the IoT, metaverse, and AI advances, 3D printing apps will branch out even more. So, I'll open the challenge to you to produce your own 3D printing app.

Social Media

There are communities of 3D printers with social media pages and groups on YouTube, Facebook, Twitter, LinkedIn, Instagram, and Meetup, etc with access to sales, demonstrations, interviews, conventions, exhibitions, events, and in-person meetings. However, such gatherings have not translated into general cultural recognition or sales outside of the industry.

I did look up Meetup groups specifically for 3D printing. There were over 630 groups listed worldwide with almost 400,000 members. In the UK there were 16 such groups with over 12,000 members, and in the United States, 16 groups have over 7500 members. With high numbers of organized and informal groups and memberships across the other online sites, it would be hard to present a coherent 3D printing front. Most of the groups have differing specialties and agendas. However, as part of two different types of Meetups groups myself, I have experienced how such communities can have insular views and practices when not engaging with the outside world in their daily lives. And since I have not joined a 3D printing group (my bad!) it would be interesting to see how they operate and spread the 3D printing message beyond the print-and-sell ideology and sharing printer stories and technology advances.

It's hard enough trying to get noticed within a niche industry, but trying to penetrate the general public's bubble and garner interest would take a gargantuan effort that only elite businessmen, celebrities, sports personalities, or social media influencers can provide. And there's your 'in'. Try to get an endorsement or appearance by one of the above. Don't pester or bombard them with 3D printed gifts, but try to establish a genuine connection where 3D printing would add value to their life and, importantly, to the lives of their thousands or millions of followers. That is, after all, the cultural goal of 3D printing.

All it would take is a little imagination and social media could open up avenues of revenue and influence to take 3D printing into the consciousness of the public and become a viral commodity.

There is also another big positive about social media within the 3D printing community (at least so far that I've seen on the more popular sites). There's no toxicity and tribalism. I'm sure there may be disagreements over printer types and materials, etc., but unlike other cultural realms which seem to generate constant anti-social media, the 3D printing community tends to help each other and engender innovation. And that's another cultural virtue.

AI Advertising

From 15th to 18th November 2021, I attended the virtual event Silicon Valley Comes to UK (SVC2UK). One of the talks discussed the intersection of AI and advertising and how over the past three decades technology has changed the way we access information. The traditional media methods (let's call it H.I. - Human Intelligence methods) above have been surpassed by the seemingly limitless digital content which has created a so-called 'attention economy'. You're no longer competing against other companies for the audience's attention, but also just to be noticed at all.

But will AI change this? And can 3D printing capitalize on its potential? Remember the internet is not free for content producers, it's funded by advertising, so why not invest in an AI advertising platform or service? AI advertising could make predictions based on your customer's purchasing habits and production processes. Your decision-making can also let your AI create a digital-self of you and/or your customers for data modeling, learning, and growing in order to best serve your needs, capture more customer engagement, and cater to an audience that will have their attention focused on you.

An advantage would be to have your advertising program's results measured, customized, and perfected to simplify and automate the process. One way to do so may be through the metaverse whether the new open internet is free or a paid-for service. Having 3D printing content on the metaverse steered by your AI advertising will ensure the dissemination of the message across new mediums to your dispersed markets.

Market. Message. Medium. That is the core of advertising. Once 3D printing harnesses these three attributes, with the right marketing strategy, targeting the right people with the right message, and through using the best medium, 3D printing will be on the cultural map.

3D Printing as a Messaging Platform

What? I hear you shout. What fresh blue-sky thinking is this? But bear with me. If a picture is worth a thousand words, then what could a 3D-printed form communicate? Can we make 3D printing into an encrypted messaging service? Like 3D faxing, but with physically-encoded prints. Let's call it a sculptagram. How would it work? Well, you would have an interface between your mobile/ computer and 3D printer, like an AI controlling a new type of G-code/slicer combination, utilizing a predetermined cipher for messaging. You could then email or text that interface a message (including images) which would encode it as per your predetermined cipher, slice the encrypted pattern, and send the file to another 3D printer. The receiving printer would have its own cipher equipment to decode the message to print. The message would only be able to be read once printed, perhaps with the use of a 3D scanner or using an AI program. This would keep messages private as without the cipher a person would just see a 3D-printed object without knowing its form contained a message.

Think of the Maya glyph writing system or other image-based writing forms, known as logosyllabic scripts. Each image formed a concept, or a word or part of a word in a sentence. Deciphering such scripts has been difficult and there are still other Mesoamerican scripts to be deciphered due to their complexity. So logosyllabic-type messaging would be a new 3D printing industry with complex messages able to be sent. What would such a message look like? Well, whatever you wanted. So, perhaps a thirty-centimeter by five-centimeter vertical wave-form with small slits and serrated edges conveys one message, but leaving out the gaps and edges and incorporating embossed reliefs conveys another. The form, slits, and edges are the coded message unable to be read without the cipher. And putting multiple forms together could create more complex messages. Add in 4D and 5D printing and you'd have physical messages that could also vary size and shape over time or be environmentally adaptable.

To be honest, I'm not sure on all the actual practical applications, but it would be a useful tool for communicating remotely in a 3D format, especially for puzzles, gaming, AI learning, and metaverse applications. Perhaps it can also be used for the visually impaired using 3D printers as a tactile message system or for translation purposes. At the root is the formation of a new 3D printing function beyond a printing and selling technology. Let's call the service something catchy like 3ssenger or S3nder.

Once you master the media through which you intend to proselytize the merits of 3D printing, you can take it on the road to events and network, whether at physical events, online, or as a client or a customer.

CHAPTER EIGHT

3D Printing Events & Networking

When I was setting up my company, I had to decide how to market it and grow a customer base. I'm an introvert really. I write better than I speak, I believe. Sure, when I'm with my fellow geeks and feeling comfortable I can give a quiz and be the life of the party, but speaking to strangers about myself and my work, not so much.

It's hard putting yourself out there validating yourself and your vision, and appearing confident in your work and ambitions. I had quit a decent job to start my own business with no guarantee of success. But I had the belief I could do so. I just needed to get the right pieces into place, meet the right people, and implement my plans.

So, I targeted various tech events to attend in 2021, whether virtually or in person. I was looking for ways to connect 3D printing to other technologies and communities. I believed that the business culture of 3D printing had to evolve in order to grow. Below are the blogs I wrote for the events on my website, at the time.

IoT (Internet of Things) Tech Expo

September 6th-7th 2021

"I'm here at the IoT Tech Expo, London, learning about IoT, 5G, cryptocurrency, cybersecurity, and AI, and how they can integrate with business. Of course, my interest is how we can integrate some of this with 3D printing.

Some of you already do so by remote printing or viewing your prints remotely, but what if we had smart 3D printers able to order parts like filament, nozzles, bed mats themselves or to alert you to an issue?

Apart from home use, this would be great for large corporations, schools, hospitals and business innovation. Data would be collected to log and show best print outputs and add value to your workflow.

While I asked a few IoT companies about integration with 3D printers they didn't have answers. Why not? That's what 3djacent is for, to ask these questions then throw them out to you for discussion, innovation, collaboration, and hopefully solutions."

I wrote the above quickly and excitedly in my notepad on the second day of the conference after listening to almost a dozen panels and talks over the two days. It was unseasonably hot for September, so I chose to make the short vlog from my notes inside on the upper level, for my YouTube channel.

It's funny that while IoT companies like Nivid and DataShurubs told me they offered 3D printing services they did not integrate this with their IoT services. Yet after the conference, I attended an entrepreneurs' meetup and one of the attendees told me he worked for Hewlett Packard and that they have 3D printers that do alert them to the fact they require new filament.

So, what is the state of IoT within the 3D print industry? I would hazard a guess that it's at two extremes. The first is where large corporations are using IoT sensors and networks for monitoring and fault diagnosis and the second where small companies or individuals have to use a camera/app set-up to remote view prints and alert them to any issues on their smartphone. But can we do better?

With so many of us working from home can the 3D printing industry benefit from this? Online sales of products skyrocketed during the lockdowns and will probably be the norm. With so much being bought, it's forgotten that so much could have been 3D printed at cheaper costs. This is the time for us printers to be advertising 3D printing courses, printers, products, services, and apps. This is the time to get printers into domestic settings or at the very least advertising, even on TV, marketing the enormous benefits of 3D printing. This is where we should be developing a plug & play printer for homes, with a catalog of products on an iPad or other tablet device for people to order their print to be delivered or printed at home. And it would have been ordered through an IoT service.

As I have stated before, people understand the internet. They can work a TV, microwave, and X-Box without having to take a course in it. So why not a 3D printer? The tech is there to make it possible. And IoT services would make it even easier with AI taking over a lot of the demanding user interface issues.

At the IoT conference, one of the panelists stated, "Begin with the why of the tech, what value would it add? Then search for the tech to use." Too often we develop a tech and try to apply it to many situations or to solve a problem that's not there.

3D printing can solve the 'Why'. 3D printing technology is already resolving production design and cost issues, creating new aesthetics, and preventing resource wastage. We need to be more proactive in merging 3D printing with IoT and other emerging technologies to enhance the capabilities of the industry. By the time 3D printing hits its 50th anniversary in 2033, the 3D printing landscape will be vastly different, thanks to IoT.

London Tech Week – Virtual

Monday 20th to Friday 24th September 2021

Urgh, my head hurts! I was at London Tech Week (virtually at least) learning about various industries of tomorrow. I watched and participated in panels on innovation, learned for the first time about CreaTech (creativity with technology), heard about the coming of 6G, and made some good connections. Then came the sprawling idea of the metaverse.

So, what has the above got to do with 3D printing? Well, everything. As with IoT, the UK market is opening to new and innovative tech companies. The ability to meet new business partners and adjacent communities, collaborate with competitors, communicate more intelligently with our customers and gain more referrals would be enhanced. How? With more people online than ever before innovation in marketing strategies would create better support organizations, but also include emotional and functional human engagement. Our creativity in 3D printing is a way industry tech can affect a different and better outcome for ourselves and customers.

CreaTech

I had never heard of CreaTech before—creative industries and technology helping to scale products and services for CIC—cross-industry collaboration. But the creativity part is somewhat under threat. The creative fields in Arts & Humanities drive CreaTech industries, but government policies are cutting back A&H budgets. This highlights their ignorance that creatives manifest within business circles in unstructured and nuanced ways. Creativity is a complex field to quantify and government market policies and growth targets are based on the economics of salary and not skills. Creatives slip through the cracks even though our creative application of knowledge and entrepreneurial spirit drives much of the growth in the industry.

And so with 3D printing. CreaTech will be more immersive with new ways of telling and sharing stories. 3D printing is at the convergence between art form types. We will be using 3D printers in both the real and digital landscape, democratizing manufacturing, and creating new games and educational areas. And just like FinTech, MedTech, and EdTech, CreaTech will break out of its niche and spread its benefits to other sectors with 3D printing as a central pillar.

Cryptocurrency

Let's take in cryptocurrency and 3D printing. Just as with manufacturing being disrupted by 3D printing, so cryptocurrency or programmable money is disrupting the traditional currency economy. However, it is still at the relatively primitive stage trading in mostly highly fluid ideas like GIFs, art, media, or the meme economy. Sure, at the moment we can use crypto to pay for 3D-printed products, however, in the future crypto won't replace money. It will morph into something different allowing more transparency and security with future assets tokenized as digital and creative media. 3D printing could help bridge the gap by producing its own tokenized economy. Perhaps iconic or innovative STL's, designs, or other 3D scanning augmented reality art or programs could be included in the NFT class. On the physical side, I have further ideas to link 3D printing to cryptomining, which will be presented in Chapter Nine.

Technovation

3D printing is definitely in the 'technovation' arena developing new ideas, products, and services through technology. We saw how 3D printing made headlines during the worst of the pandemic with individuals and companies printing masks, faceshield parts, ventilators, and other medical products. People became more creative in the face of lockdowns and redundancies; they told their own stories and communicated their creativity. But at the same time, there will have to be a strategic automation of complex jobs, so masses of employees are not precluded from working.

With the pace of tech rising even through the COVID-19 pandemic, tech investment in FinTech, EdTech, MedTech, etc., will have to offer a way for people to upskill. CreaTech would be an obvious outlet. An issue that was noted, however, was tech's proclivity to promote loneliness, especially on social media platforms. The basis for this is that for some people social media elicits a dopamine reaction, a short-term pleasurable effect (e.g. waiting for instant likes and other reactions) rather than an oxytocin reaction which is related to empathy, trust, and relationship building, a long-term feeling of wellness (e.g. conversations and thoughtful messaging). So no matter how far we go in tech, in-person interaction is still the healthiest option. With 3D printing we can offer that tactile sense of interaction, the human-made artifact, and provide that connection between tech and humans.

Metaverse

Then came the Metaverse! The so-called new internet is described as persistent immersive, shared, decentralized community/social spaces. If you've been to a Fortnite concert or played in VR/AR games, or experienced a 3D simulated space or experienced crypto/NFT you've been in the Metaverse. It's not a joined-up community or movement yet, but it's on the way. It's the gamification of content to engage customers. You'll be soon meeting, not on Zoom or Teams, but as avatars inhabiting a room. (I actually witnessed this while attending the 2021 Mars Society annual conference. They had a virtual room with avatars and they were fully interactive even asking questions to the speakers with their avatars).

The 'default' world (real world) will give way to an increasingly digital landscape using avatars and exponential technology run on game engines. So your online persona may become more important than your real-life counterpart. Weird huh? [Author note: at the time of writing this original blog, I repeated what the speaker stated about Fortnite being a metaverse game. There seems to be a heated debate as to if it actually is. However, metaverse games include: Decentraland, Axie Infinity, Sandbox, Alien Worlds, Second Life, Illuvium, Roblox, My Neighbor Alice, Minecraft, Voxels, and Battle Infinity, and more.] So where would that leave 3D printers? Well, as I learned as well, we are in the world of digital twins—digital copies of real-world products, buildings, and spaces, etc. We can have digital twins of 3D printers, prototypes or running digital sims with established printers. Want to run a 3D printer test on the moon or Mars? It will be tested in the metaverse. Want to play with a printer across the world? Then try a virtual one. The teaching of 3D printing could be easier in the metaverse, with hundreds of student avatars from remote locations tuned into an interactive class. Remember how Tony Stark in the movie *Iron Man* could split apart, rotate, reassemble, and manipulate holographic parts of his armor before printing it while interfacing with Jarvis, his AI butler? In the future we'll have such holographic capabilities, transferring those holographic files to be 3D printed either in the metaverse first for proofing and/or sharing with others or straight to the real-world printer. Let's just say the 3D printerverse is coming. And how would that work? Well, I'll just invent metaprinting, as described in Chapter Nine.

6G

The future communications network was also discussed. While it is still in the academic and research stage, it has the potential to create emergent and/or convergent disruptive technologies. The big ask is to 'throttle down' on the overhyped overselling marketing aspects of massive multiplayer role-playing games and multimedia as was the case with 5G. Expectations will have to be managed to overcome resistance to 6G. It cannot be rushed and its execution will have to include global standards and visions with well-defined back-end business plans, spectrum band usage, end-to-end responsibility, and the exponential platforms and infrastructure to work out issues. There will be necessary different frequency ecosystems (e.g. verticals for differing industries like shipping and aircraft). And uses will most likely be rolled out to satellites, holography, AI, software architecture, flexible networks, machine learning, organic networks, smart networks and sensors. It should boast a better sustainable development model being more energy efficient (e.g. less carbon intensive) and affordable. And of course, 6G will be prominent in the forthcoming metaverse and virtual domains.

And this is where 3D printing would be involved as 3D printing will be in use on the moon and Mars, making rockets and satellites in space (the ISS has had a 3D printer since 2014) where connectivity to services would be vital. There'll be virtual 3D printers whether in games or simulations, holographic printers, and AI-run 3D printing farms. There was further talk of 6G use for quantum computer use and the emergence of 7G, but then my brain started to melt with all the data input and my burgeoning ideas for the future of 3D printing. Needless to say, 3D printing will be more adaptive in the IoT world under 6G connectivity.

And that was a week of tech and how 3D printing can be heavily involved with the upcoming wave of new creative and exponential technologies. We're ready for it!

Paris 3D Print Congress and Exhibition

October 20-21/2021

I officially launched 3djacent in Paris. Why? Because everything came together at the same time from the timing of the event, my products available in the shop, the website going live, and my goody bags ready to be handed out to vendors I spoke with. I had been to Paris a couple of times and was familiar with the metro system, but I stayed at a hotel just around the corner from the venue, the Palais des Congrès de Paris.

Once I got my goody bags filled on my hotel room bed, stuffed with branded business cards, bookmarks, pens, A5 flyers with QR codes, and most importantly little chocolates and sweets, I was off to the races. The Palais des Congrès is a massive space, an odd-shaped, sharp-edged concrete box serving a purpose akin to London's 02 or Excel center with retail units, restaurants, cinema, and exhibition spaces. After being checked for my ticket and Covid pass, I wasn't sure where to start, but I bucked up my courage with my carrier bag full of smaller goody bags and talked to the first exhibitor. I wasn't selling anything physical, but I wanted to establish a connection, to build up my credibility within myself and with each exhibitor. I asked what they did, how they did it, and other things like the advantages of their system or product over others. I'd then relate what I was doing as a 3D printing community advocate and how they could sign up in our directory or contact me for reseller and affiliate options. Such conversations then led to other personable chats such as finding out one of the French vendors actually lived in west London. *Quelle surprise!*

I was really impressed with the technology on show from manufacturers; the different types of printing equipment and accessories, scanners, materials, post-processing options, holography, 3D printed products, and the people. Sure, I had seen 3D printers before, but the variety here was mind-boggling. I guess I had previously only concentrated on domestic options and on plastic filaments, but more 'advanced' (in my eyes) setups were far more interesting with resins, metals, ceramics, powders, and pellets on show. Each company I spoke with got a goody bag.

Companies I spoke with included Cubeek3d with Ultimaker products; e-Motion Tech who were busy printing up a guitar frame and also had a dual-headed printer; CETIM; M4P; Smart Metal Design with chairs, a table stand, bottles, and bags produced through their electroforming process; Dagoma 3D; Alsima 3D industries; e-Nable France with their plastic hand prosthetics, Boreal; Lemantek; North Maker who were sensationally going for the world record with the most amount of 3D printed model train carriages measuring over 10 meters. They wanted to retake the record from the Swiss. This was to be at 1pm on the Thursday.

Millennium 3D had a super cool video showing off their products and processes; Markforged displayed sintered products and multiple materials; joke Eneska had a large postprocessing platform with sanders, polishers, deburrers and powder removal tools; 3ntr; Stratasys; Artec with their own and Scantech portable scanners. And while I saw a rep from A3DM magazine, I didn't get to speak with him as he was busy. And there were many more I didn't get around to. I finished the first day just after 1pm as I was running out of goody bags, and I had more time the following day. So, off I went sight-seeing and to reflect on a great first day.

One thing I hadn't done was to join in on workshops or watch the start-ups competition. I don't speak much French and the exhibitors were very kind to speak English back to me, but the exhibition was predominantly a Francophile affair. I had no doubt that had the Pandemic not occurred then more continental, British, US, and Chinese companies would have taken part as many of the companies I spoke to either had parent companies in those countries or partner companies in the UK, I was told I could contact. So my goal was to introduce myself, network and market myself, and see what the industry was like outside of my own bubble. I had a lot of thinking to do about the nature and trajectory of my start-up.

The next day, I was met by a good friend, a native Parisian, new to the world of 3D printing but eager to learn and to translate for me. She read my flyer material, took goody bags for me, and jumped right in explaining/translating in French who I was and my business. She was really great, talked a lot, and was well-received. She was also amazed by the 3D printing equipment on offer and what they could print, especially the shoes, bikes, and metal products she saw. We met with Altair who in turn gifted us with their own branded webcam cover, mouse mat, and brochure; and while I met with ValorYeu who turned fishing nets into filament, they didn't take any of our offerings as they had to recycle everything they had. And stupidly at the time, I forgot I had my digital business card on my phone.

There was a fellow visitor nearby who 3D printed boomerangs and was going to hold a boomerang competition in Paris for 3D printed boomerangs. Voluminous was one of several companies that used 3D printed components for their printer's frame; Tundra with their large Hexadrone displayed by

Sculpteo part of BASG; F3DF/Autodesk with their demonstration piece that looked like a mechanical demigorgan; Viaccess Orca; Delta Equipment; TCN who produced dyes for 3D printed materials; TH industries; and I finally got to speak with the rep from A3DM magazine, who liked my idea and was quite helpful with advice on investment opportunities.

Midway through the day, we got our spots to witness North Maker's world record attempt for 3D printed model train carriages of which they had sixty. After announcing their intention to the gathered crowd, an official went around and measured the 10m distance and counted the 60 carriages. I recorded the sequence for posterity myself. And they did it, a successful 10m run. I expected them to go as far as they could, but it seems 10m was the set target limit. After, the official then physically inspected the carriages to make sure all were 3D printed. I hope to see more records like this attempted and publicized more so that 'ordinary' people can gain some appreciation and experience for 3D printing, especially children.

We networked again, working the floor, re-meeting vendors for a further chat, and in all gave away 46 of the 50 goody bags we had, which was good. Some of the vendors were now packing to leave, so we bid farewell to the Paris 3D Print Congress. For me, it was a very successful exhibition, eye-opening, and with prospects to pursue as leads. Once I got back to London, I would be following up on all the contacts I had made.

With everything I had seen and learned, I now truly believe there is the technology to be able to create a plug & play 3D printer for domestic use. I've touched on this before and will continue to do so because industries, workforces, and society will continue to change and so will the need for alternative services and manufacturing. The question is as always: is there the will to do so? The market is there and we are on the cusp of revealing 3D printing's real potential for change. And such exhibitions will surely be the breeding ground for challenging that change. It was for me.

The exhibition is usually held in Lyon and will be again in April 2022. Now that I know what to expect, with the experience I gained, and hopefully, with the growth of 3djacent, I am considering attending. Thank you, Paris.

Silicon Valley Comes to UK (SVC2UK) - Virtual

Monday 15th to Thursday 18th November, 2021

Unicorns and Zebras

While this was not exactly tied to a 3D printing agenda, I still learned a few things that could be applied to 3D printing. The first part of the conference concentrated on aspects of scaling one's business and dealing with the virtuous circle while avoiding investment bust cycles by investing, growing, exiting, and reinvesting the funds.

There were trends to look out for, especially during Covid where everybody was now on the same level playing field. One key thing I could connect to 3D printing was a similar outlook to Peter Diamandis' 6 D's with a 4 D list focusing on Disruption, Digital Transformation, Distributed Workforce, and Data.

3D printing will bring disruption with changing staff expectations and workplace evolution; digital transformation (DX) will be more mobile and on mobile and cloud platforms with simplification and automation; while a distributed workforce won't be working in a hybrid mode for long term but learning to manage churn, skill shortages, wellness, and remote culture, creating a sustainable work-life balance; and lastly data will require new types of interfaces and sharing and protection protocols. In relation to 3D printing as a disruptive technology, it will certainly be a future leader in entrepreneurship, challenging the status quo in the long-term industry trends.

Part of 3D printing's potential and legacy is in the line between Unicorn (tech disruptors) and Zebra (adaptive, inclusive, ethical community) industries. Unicorn businesses include Facebook, Apple, Google, TikTok, and Amazon, etc., which hold a huge sway over our lives with a zeal for monopolization and huge profits. Zebras on the other hand want to nurture that entrepreneurial spirit with sustainability, quality, and morally-sought profits. Such companies include Borderless Global, Patreon, Toya, and Zapier, etc. 3D printing isn't a unicorn but is part of that exponential world. However, it can be part of the zebra community ushering in that value-orientated society.

In one of the lectures, Michael Hayman MBE, co-founder of communications consultancy Seven Hills put it plainly that we are both the first generation to realize we are destroying the planet and the last generation who may be able to save it. We need to leverage technology to benefit the world with a digital future filled with purpose and compassion. I believe 3D printing will also be part of that future solution, which I will discuss in Chapter Ten.

The future of entrepreneurship was discussed and the eco-benefits that new kinds of working practices, especially zebras will bring. Over $1Bn is already being spent on people and companies involved with world-changing environmental, climate, and sustainability plans. Some of these are long-term goals which could potentially deter investors not patient enough for returns to back long-term businesses that have a 5-to-10-year plan. It has been predicted that the 2020s will prove essential for environmental, social, and economic prosperity. By 2031, the size of companies will change as cities evolve into a digital phase with construction, transport, health, food, and clothing transforming cities with crypto, DeFi, and automation bringing more flexibility in the industry. By 2050 there will be 9 billion people to feed, clothe, transport, employ, and educate so flexibility has to be key to our survival. Unlike any generation before in history, we know how to design and build for the future.

The new mantra is to go fast and fix things on the go rather than the old break things and build anew, circling back to zebras and unicorns, respectively. We need 3D printing to move beyond its unicorn role as a disruptor of manufacturing and grow into an actual world-changing producer with the capabilities and experiences as a zebra.

But even better, years after this event, I recently learned of a third entrepreneurial path, that of phoenixes; companies that are proactive in regeneration. These are massive enterprises contributing billions of dollars into protecting and restoring ecosystems. And as described in this book, 3D printing recycling measures and manufacturing techniques can go a long way into helping its own industry and others cultivate a culture of sustainability and regeneration.

Business Groups

In 2019, I had also joined a business Meetup group, Founder Nation (formerly Entrepreneurs in London) founded and run by Patrick M. Powers and his team. It's a great group for meeting up with other business folk, networking, learning, amid a few drinks at the various pubs on a monthly basis. We had ice-breaking games, various giveaways, and sometimes separate lectures by guest entrepreneurs. While there were a lot of accountants at these meetings, I did finally have to use one of their services for my corporation tax and Companies House annual accounts. There are plenty of other business groups out there and I would recommend joining one for the experience and knowledge gained. This one had low ticket prices for events, was held in pubs, and you just paid for your own food and drink.

It was here that I formed the notion for my company, had the idea for this book, and where I was introduced to my proposed business partner, which would have its challenges, as detailed in Chapter Thirteen. Nevertheless, I found the nights we all met up a great ideas incubator and fountain of opportunity.

Investing in 3D Printing

Invest in the investor, bank on the bank, I say. I had quit my job in July 2021 and hoped I had enough to survive on my company's expected profits alone. I had started a Wix website, which I found out would not be up to the job I required for the interactivity I needed for the site. So, I needed more money. I looked to stocks and cryptocurrency as one way, besides my writing skills as a freelancer.

I joined a trading site and a crypto site and invested in unicorn companies like Apple, Google, Tesla, Amazon, and Facebook. I found that while others might not want to invest in such companies due to widely reported unfavorable tax and work conditions, I decided to invest in them to make money, withdraw and use it to do good myself. I didn't feel hypocritical over it. Those companies aren't going anywhere for a while so I would use their capital to increase mine for my own benefit. I didn't have the luxury of time or the money to feel otherwise.

I was surprised to find a few 3D printing companies trading so I invested in them and related industries. I also wanted to start mining cryptocurrency (more of which later) for a specific research reason other than just minting bitcoins. However, this initiative didn't last too long. While I am not actively trading, I still have the accounts and experience, which I hope to return to one day. With the potential of 3D printing growing exponentially, it may be prudent to purchase stocks in the 3D printing industry and make some money either passively or engaging in a world-changing technology.

Strangely enough, in February 2023, I saw a Facebook post by Yes, That's 3D Printed entitled "Here's a pro tip: NEVER invest into 3D printing stocks" which focused on screen grabs of investment in 3D printing tech and how stocks had plummeted over 5 years. As of February 2023, these were the stock prices researched from the Nasdaq market summary. I have rounded the numbers for clarity:

Company	2018 value (unless otherwise stated)	2023 value	Percentage loss/gain
Legacy FSRD Inc	$10 (2021)	<$1	-99.S%
Shapeways Holdings Inc	$10 (2019)	<$1	-93%
Organovo Holdings Inc	$21	$2	-90%
Markforged Holding Corp	$10 (2020)	$1	-88%
VoxeljetAG	$22	$3	-88%
Desktop Metal	$10 (2019)	$2	-82%
Velo3D Inc	$11 (2021)	$3	-70%
Proto Labs Inc	$113	$37	-68%
Nano Dimension	$7 (2021)	$3	-67%
XometryInc	$70 (2021)	$35	-50%
SLM Solutions Group AG	$38	$19	-50%
Stratasys Ltd	$20	$13	-34%
Prodways Group SA	$4.50	$3	-33%
Materialise NV	$12	$10	-16%
30 Systems Corp	$10	$10	+5%

Only 3D Systems Corp seemed to buck the trend over five years with a 5% gain. While the title conjured a negative feeling over 3D printing, my view was that it highlighted the overall volatility of stocks in general, of tech stock in particular, and in the unfamiliarity and low perception of 3D printing to non-industry traders. And of course, some of the stocks hit highs during the Pandemic and have subsequently fallen sharply. Some of the comments on the post surprisingly reflected that gloomy outlook though there was some positivity in support of 3D printing investment. I sent in a reply. . .

> This is quite interesting. I'm writing a book about some future aspects of 3D printing and one of my topics was investing in 3D printing which is a smart idea. To make a long story short, if I didn't have to pull my money out of stocks for personal circumstances, I would have kept the money in for the long haul (despite any losses). The current 3D printing technology will advance and more publicity will bring 'non-industry' interest in lifting share prices. 3D printing is still in its nascent, disruptive phase, like crypto, metaverse, etc, but early adopters can see the potential and exponential possibilities for the future. Not all companies will rise and new ones will fill the vacuum, but personally investing in 3D printing is a no-brainer for me.

Yes, That's 3D Printed did get back to me with an interest in reading this book when published, which I gratefully accepted.

Stock trading represents confidence in the product, the company, and its upward trajectory. And of course, this doesn't just pertain to investing in companies that manufacture and sell 3D printers and equipment, but also those non-3D-printing industry companies that use 3D printers, their applications, services, and technology to upgrade their own services and products beyond traditional manufacturing means.

As with 3D printing media, we need more visibility of 3D printing companies in the markets demonstrating what they can achieve. There may be pushback from traditional industries competing against 3D printing, but in the end, technology and industries evolve and 3D printing may be left standing among the top tiers of the new Industrial Revolution. The verdict on 3D printing's development and future has not been fully settled, but its investment potential is well worth the while as there is no sign of the technology slowing down.

The main variable may be the shift from large corporations (quasi-unicorns) to smaller more agile companies (zebras) sharing ideas or open-sourcing their equipment and product designs so improvements can be made, enabling sustainable 3D printing (phoenixes) to mature. As more decentralized 3D printing services arise, investment will be more personalized. Most likely, the next big 3D printing advancement will be from someone's project in their garage rather than a corporation.

The last investable interest was something I was keen on and even had a 3D printing crowd-funding category page on my website dedicated to investing in start-ups or finding investment, whether 3D printing Kickstarters or other platforms, as described below on my website:

> Looking to invest in the latest 3D printing project but not sure where to find them? Be as inventive as the inventor and search out campaigns on crowd-funding sites. You could unearth the next 3D printing gem, polish it, and earn yourself a mint while nurturing new talent.

> Have a fantastic prototype you want to share and get funding for? Promote your work on crowd-funding platforms with a campaign to seek support and investors. A successful campaign could accelerate your start-up and an exciting future.

3D PRINTING EVENTS & NETWORKING

Not sure how to get started in crowd-funding? Get tips and insights into the growing market on new projects and start-ups, and understand the successes and pitfalls before starting a campaign. Check out the linked sites or reach out directly in our forum and blogs.

In November 2022, I invested in two 3D printing campaigns on Kickstarter; a new food extruder and a new type of printer. They were just small $10 pledges, but I wanted 3djacent to start off a trend in 3D printers supporting each other and building an investment network with dedicated platforms, links, and contacts.

Maybe that will happen one day, if it hasn't already elsewhere. Sharing and trading are cultural acts, even if digitized. So, with the growing popularity in 3D printing, we have to network with and invest in ourselves and the industry. And of course, I'm open to investment in my ideas to direct 3D printing's growth through our shared culture.

3D Printing Online Games

I briefly discussed 3D printing's role in the metaverse and with 6G, but what about the gaming world which will be an integral part of networking and eventing within the metaverse? The gamification of 3D printing to advance domestic and commercial uptake would also increase the technological innovation of the industry.

3D printers could be part of gaming worlds such as Sims with realistic depictions of 3D printers providing another social interaction. Maybe you could even earn money from virtual 3D printing or interacting with other participants to earn money in the real world. War and adventure gaming sites could have links to licensed sources, external companies, and individuals who can 3D print figures from the games, earning them all extra revenue. Perhaps using a metaverse platform like Earth2 will also provide an avenue for 3D printing use.

Through Earth2, you can buy your land in the metaverse and when visualizing your buildings, you can 3D print scale models for the real world. You can test out architectural models and as the platform is an exact topographical representation of the world, you can show prospective clients and fans what the buildings would look like in a natural setting.

Gamifying 3D printing wouldn't just be about education, but also incorporate action and adventure. With my humble abilities, as I am not a gamer or game designer, I had come up with a simple 3D printing game concept called 3D Printectors of the World (Printectors a play on the word Protectors):

3D Printectors of the World

Premise:
Aliens have invaded Earth intent upon destruction and domination. The only salvation are teams around the world 3D printing equipment and weapons to neutralize the aliens.

Game type:
1st-person shooter/ multi-role playing game (You can be a soldier, scientist, or other player). Free sample play then subscription fees apply for full game.

Game story:
You and a group of soldiers and scientists are tasked with finding a huge underground secret bunker in a network of tunnels and labs where a revolutionary 3D printer awaits in the deepest level. It will print anything you want (weapons, spaceships, and equipment) to defend Earth.

The aliens are tracking you. You only have a set amount of hours to find the printer, set it up (i.e. choose settings, bed level, choose materials, heat bed, etc), and select what to print using the touchscreen on the in-game printer console, how many items, and how large, though such decisions will sacrifice time in which to defend the world. The aliens can break through at any time.

While you wait for the print to finish (which can be hours in real time), you have to defend the printer and lab against the aliens who have penetrated the bunker and want to destroy the printer.

Outcomes:

If you lose (i.e. the aliens kill you or destroy the printer or the print fails) you have to start the game or level again.

If you win, the printed items can be used by your team against the aliens or 'shipped out' to other online teams.

As an added bonus, whatever you complete printing in the game can be saved to a file, printed as a scale model in the real world, or shared with others.

******

How's that for a game? I'm sure you can do better and incorporate 3D printing into it. The controls on the in-game printer would be accurate in regards to 3D printers so you are actually learning the control set-ups on a printer and what it takes to heat the bed and choose what you want printed and the materials to be used. And most importantly, the in-game print time can be adjusted for hours or days and the game paused so you don't lose your place. You wouldn't want to do so when printing up a spacecraft over three days now, would you? Challenging enough for you?

Another concept I'll call 'Hero Maker' could involve you printing up an in-game character's superhero costume to fight with and against other players who have also printed costumes for their characters. There could be competitions for best costume and maybe some could be printed in the real world for cosplay events.

Of course, other games will have different storylines and points of learning but having games similar to others will make it easier to comprehend and adapt to new technology. However, you could gamify 3D printing recycling, compete to build new types of 3D printers and develop new print materials, and even play games to win STL files and new 3D printing NFTs.

The gaming industry is set for another technological leap in the metaverse, so you can imagine putting on your goggles or VR headset and actually be in the game, 'touching' 3D printer controls and feeling, seeing, hearing the workings of the machine. You could connect remotely to 3D printers worldwide and have (virtual v3D or electronic e3D) 3D printing competitions like the e-sports revolution. Such techniques would come in handy for virtual learning as well, so courses wouldn't be just online, but also have a hands-on virtual element as well. E-learning will take a backstep to V-learning and V-gaming as virtual 3D printing (v3D-printing) takes over.

So, you don't have to 3D print in your home, you can virtually be on a beach in Hawaii printing away or on the moon. You don't have to be alone and can be part of a 3D printing farm underwater or in the middle of a football stadium. And what you print virtually will be copied onto a real printer anywhere you want. The metaverse, gaming, learning, and 3D printing are going to coalesce and elevate what humanity can achieve in ambition and creativity.

Events and networking occasions such as those described above will be the lifeblood of the 3D printing culture. The events don't have to be purely about 3D printing. Such diverse technological and social events will breed new ideas, cross-pollinating creativity, and a multidisciplinary approach will advance 3D printing in other cultural milieus. If not, then 3D printing will stagnate and be bereft of creative and positive influences. We don't want 3D printing falling prey to corporate and industrial gatekeepers and ideologues wanting 3D printing to stay in some pure technological arena for them and their own financial gain and patented discoveries. Culture changes, and we need the 3D printing culture to realize its potential and render something new for the world.

CHAPTER NINE

Exponential 3D Printing

3D printing is part of the exponential wave of new technologies such as cryptocurrency, AI, nanotechnology, robotics, the metaverse, and 6G on the way. Exponential technology describes a state where its growth is not linear, but exceeds a rate of pace that disrupts economic markets and positively affects lifestyles; exactly what our 3D printing cultural journey requires. The first real exponential was computers with its rate of growth measured by 'Moore's Law' which stated that computing power would double every two years. Computer power may be hitting the wall in a few years until quantum computing arrives, but other technologies have their own growth rate 'laws' and the faster the growth the faster the rate of the technology develops.

But while 3D printing is an exponential technology on its own, it can be combined with other exponentials which are already or will be a part of our cultural lives to greatly exceed its own self-advancement. So, let's see where such ideas could lead.

4D Printing (Adaptive 3D Printing)

The 4D printing process allows a 3D-printed object to transform itself into another shape or structure via the input of an external energy source such as temperature, light, pressure or other environmental forces. The 4D adds a time element into the equation just as 3D adds a volumetric advantage over 2D objects. Think of a flower petal opening and closing over time in response to external influences such as insects, rain, sunlight, etc and 4D printing has a similar function.

An advantage would be to have self-assembling and programmable material enabling technologies to enhance, reinvent, and repurpose services such as product design, manufacturing, construction, and performance. 4D printing will also lead to new advanced and smart programmable materials. The basis for the new materials could be hydrogel in nature or a shape memory polymer, with thermomechanical and other smart properties. This will enable shape, size, and structural changes (such as folding) creating a big advantage over rigid 3D printing materials.

The potential for 4D printing is vast with transforming objects revolutionizing many industries such as:

- Medicine: Using programmable sensors to dispense remedies after detecting specific ailments.

- Construction: creating printed windows reacting to weather conditions providing shade, solar power, advertisement surfaces, and other smart useful operations.

- Space travel: building spacecraft, satellite surfaces, and sub-structures which transform to provide protection, self-repair capabilities, or anchor and docking functions without the need for extra mechanical and powered parts.

- Fashion: making smartwear and wearable technology enabling more weather-resistant and protective clothing, auto-sizing clothing for weight gain and loss, and programmable accessories for different occasions.

While the 4D printing process may not be available for household printers, many variations and applications for 4D printing technology are being researched and tested now. The benefits will be seen and experienced within the next decade, if not sooner. Take a 3D-printed car, for example, with doors sporting no hinges as it uses foldable and bendable doors and panels, or vehicles with deformable body features for varying environmental conditions, and self-repairing tires. And if that wasn't enough, let's add another 'D' to the printing equation.

5D Printing (5-axis Printing)

5D printing doesn't lead off into the 5th dimension as such, but it is printing on a 5-axis base. 3D printing operates over three linear axes: X, Y, and Z. However, 5D printing enables printing in 5 axes with rotation in the X and Y axes allowing for curved layers to be created. Due to the complexity of 5D printing and modeling design, custom-built printers are not on the market, yet. Even as engineers and companies grapple with 3D printing advantages over traditional manufacturing, 5D printing requires more years of research to consider its use in the printing ecosystem. The advantages 5D printing will have over 3D printing include:

- Making printing easier with less material used due to fewer supports.
- Stronger models with higher-quality surfaces.
- Less post-processing.

One big disadvantage involves slicing techniques as most slicers would not be able to process 5-axis printing, so new software or seeking an expert would be required.

There will be various 5D printer models, depending on which industry acquires the printer and the products to be printed, but as with variations in 3D printer models either the printhead will rotate while printing or the print bed will rotate through the additional two axes. And most likely such printers will begin life for industrial uses. But the technology will trickle down allowing for desktop models and smaller company use.

So, if you want to get ahead in the market, start investing in 4D and 5D printing, whether through researching the technology and the industries (e.g. aerospace, automotive, medical, etc.) most likely to use such printing, creating your own 4D and 5D printing models, creating new slicing methods, forming communities to be early adopters of the technology, and becoming experts before they are even on the market. And it doesn't end there. Here's another 'D'.

6D Printing (Aerial Additive Manufacturing (Aerial-AM) Printing)

Yes, believe it or not, there are 6D/6-axis printers with rotation in the Z-axis. While rare, their full potential will soon be fully realized. At Imperial College London, a team of engineers developed a type of autonomous 3D printer used under human supervision able to print structures while in mid-air. While conventional methods to 3D print buildings use either long robotic arms or a ground-based dynamic scaffold to deposit the layers of material, the 6D-printing aerial drones use a dynamic nozzle to counteract the hovering print motion to layer the materials precisely.

Imperial's process, which they call aerial additive manufacturing (Aerial-AM), functions like wasp or bee-like drones imitating the real insect aviators building their hives layer by layer in-flight. The engineers' proof-of-concept demonstration produced:

- A 2.05 meter high cylinder built of 72 layers of a rapid-curing insulation foam.
- A 0.18 meter high cylinder constructed of 28 layers of structural pseudoplastic cement material.
- A light-trail virtual dome-like print simulated by using their hovering multi-drone system to layer material all at once with three 3D-drones.

The aerial printers could be used to construct buildings faster, more efficiently, with less waste, even in less-than-accessible areas, which saves on labor costs. Their multi-drone printer system allocates each drone with a different role, one (designated as BuildDrones) depositing layers, the other drone (called ScanDrones) scanning or path-planning the structure to guide the next layer to within an accuracy of five millimeters.

Other demonstrations have been carried out by other university teams such as at Tongji University in China and institutions in Thessaloniki, Greece. The increasing amount of academic research will push the technology into the commercial realm. And as we have seen 3D printing has evolved for the better through design innovation, accessibility, knowledge sharing, and lower costs. So there should be no doubt that advanced 4D, 5D, and even 6D printing will follow this evolutionary curve and be in our homes, workplaces, and schools within a decade.

4D & 5D Printed Music and Art

So, what other innovative projects could be created with both 4D and 5D printing? Well, of course, 3D, 4D, and 5D printing will be used to create art with stunning and novel forms, but how about making art from music? There have been examples of 3D printed sounds or music visualization, notably by Gieeel in 2016 where he "captured the waveform in Audacity, did a screen capture, and then convert[ed] the image to an SVG file using Inkscape." He then used a CAD program to convert the image into a 3D object ready for printing. The image of the captured audio clip reminded me of the top part of a conch shell. It may have been a gimmick then, but 3D printing has moved on.

Imagine encoding (perhaps using an AI) recorded or real-time musical scores and performances and converting them using CAD into an 'STL' (stereolithography) file. While the song is playing the printer will sculpt a piece of structural art from the sound of the lyrics and musical notes creating a new form of art. Imagine a concert with not just a light show but also a giant 3D, 4D, or 5D printer visualizing the music, an audio performance captured as a physical form.

4D-printed musical art could change shape as different instruments, sounds, or voices are played enhancing performance and increasing audience participation. The structure would be random for each concert encapsulating individual performances.

5D-printed musical art would follow on from 4D printed, but able to complete more abstract structures with soaring curves and complex forms. Perhaps multiple print heads would be required, each programmed for different instruments or sounds, just like a symphony performance.

These would be inventive and unique merchandise revenue streams with copies of the art made whether as physical art forms, digital copies, or even as NFTs, further connecting 3D printing to other exponential tech and expanding on cultural art forms.

The main requirement for this innovative technology would be developing software able to recognize and translate musical notes and sounds into an STL file. It may be a case of inputting the written score and then having a new type of slicer visualize the pattern for printing. However, we can already transcribe voice to text, so the next step will be for music to be converted into a digital mathematical context for STLs and for slicers to perform a music sculpture in real-time. So, I would suggest you get such software patented quickly for 4D and 5D printing sculptural music.

4D and 5D printing can have enormous consequences and advantages to society as a whole. They would allow for programmable materials to be present in almost all aspects of our lives. Imagine products and items in more sophisticated 3D printed forms activated by touch, face and eye recognition, voice control, or other senses. We are on the cusp of creating cradle-to-cradle technology, able to help regenerate materials, and promote sustainability. Welcome to the art and music Revolution.

3D Printing, AI, and Rabbits

What's more cultural than language and sharing ideas, gossip, and knowledge with others? This used to be a human-to-human experience, but in the last few years, smart chatbot programs have become prevalent in our lives.

The most recent, which is seemingly taking the world by storm is ChatGPT (Chat Generative Pre-trained Transformer) launched by OpenAI in November 2022. This chatbot is part of OpenAI's large family of language models (LLM) with adaptive learning techniques to mimic human conversations. Its versatility includes writing and debugging computer programs, composing music, teleplays, and fictional tales, writing student essays, poetry and songs, and even playing games.

Among OpenAI's notable founders are Sam Altman (Y Combinator), Reid Hoffman (LinkedIn), Jessica Livingston (Y Combinator), Elon Musk (Tesla/ SpaceX), Ilya Sutskever, and Peter Thiel (PayPal), with significant investment from Microsoft. The pedigree and potential of the program is phenomenal and will lead to further iterations, competition, and advances in the field which will be transferred to other industries.

Using ChatGPT and its ilk as a 3D printing interface would be part of the missing link in creating a one-stop-shop 3D printer. As I have discussed above, the integration of AI and other exponential technologies with 3D printing will permit:

- Bespoke menu creation and interpretation of product specifications. Rather than endless online searches, an AI chatbot could narrow down and customize your 3D printing menu.

- For advertising, a chatbot, like ChatGPT, could customize your advertising, creating tailor-made advertising, marketing campaigns and revenue streams, and help engage customers 24/7.
- For creativity, a chatbot interface and software writer could adapt musical scores into 4D and 5D sculptures.

Each would consist of a conversation with the AI that would understand your requirements faster than human coding. Using ChatGPT on an industrial scale could help 3D printing realize highly-customized manufacturing processes and products. 3D printer users trying to scale up could use this as their tipping point to success.

Just as this book was going to press for the second edition, I came across an article about Santa Monica-based AI startup Rabbit which had created its own OS for its innovative r1. This mobile phone-like device uses a Large Action Model (LAM) to "understand and replicate human actions on various computer interfaces, streamlining navigation through apps effortlessly." It's like an upgrade to ChatGPT, which reacts quicker with more specific information enabling actions to be carried out and not just information gathering. ChatGPT and other large language models (LLM) allow for data collation and decision-making, while the r1, with its large action model operation (LAM), would perform the actions off the back of the information collected.

In the future, I envision such rabbits allowing you to connect with your 3D printer, whether through an app or directly, and describe what you want designed and printed. You could create images and have it g-coded. The rabbit would ask you to confirm your choice before executing your print. The Rabbit OS is far more intuitive and will have more real-world, real-time applications for activity-based technology than large language model AIs. For now, the r1 is Rabbit's main product, but imagine Rabbit systems integrated with other products, including 3D printing. And then imagine Rabbit engaged in the IoT world empowering other appliances to be truly smart. One day, you'll be having competent conversations with your 3D printer, while it prints, sends messages, and intuits your next design based on your chats.

AI is a fast-moving industry, so the next big thing could be weeks away rather than years. But it all has the potential to be used by the 3D printing industry, from grass roots to commercial to corporate. It levels the playing field. And AI will have a great impact on the next internet generation crossed with 3D printing.

Metaprinting

I'm looking at an exciting future for 3D printing in the metaverse and other alt-realities. I call it metaprinting. Why metaprinting? Because meta—Greek for 'change' —and this way of printing will transcend the real world, connect through virtual mediums, and transform the industry. Meta can also be self-referential with the virtual world reflecting the real world as the intersection between 3D printing and alt-realities. This is where you can take a design, setting, or other real world 3D printer actions and manipulate it in the metaverse, or a VR/AR medium, with added AI assistance. Metaprinting would enable real time and simultaneous prototyping of multiple versions of what you want to print, for the optimal result.

How would it work (at least in my head)? First, you would be a member of or join a metaverse site. Remember, the metaverse won't be a single joined entity and will reflect the real world with multiple organizations offering their metaverse portal, so the choice will be yours as to who you join. Secondly, you'll have VR/AR goggles or glasses or visor, headset, haptic-enabled gloves, and other sensory accoutrements for your experience. And thirdly, of course, access to a 3D printer.

So, say you want to print a mug. You design it with your size, shape, color, and decorations in mind. You may or may not be happy with it, but you decide to change some parameters of the mug or the print set up. You may choose to see how it would print with different filaments or to change the color. Changing the size would alter print time. You have a thought to make ceramic and metal versions in various colors. And then you factor in a resin print. Your choices are now mounting up with limited time and resources to print, so you go the metaprinting route.

The metaverse will enable connection of your personal devices (like your mobile, game station, laptop, etc.) to your account so you should have your 3D printer registered as a recognized device. Your print history would then already be automatically saved to your metaverse account, like a cloud service. With this you would enter your metaverse portal, select your print file, and load it to your augmented/virtual reality slicer. You can use the haptic-glove ability in your virtual slicer to 'touch' your mug to view at various angles.

But now comes the ingenious part, you can literally verbally ask your AI or alternatively haptically bring up a virtual menu and simultaneously select all the mug variations you want. The selections would be displayed holographically before you (with your VR/AR goggles) and you can inspect them and manipulate specifications. 3D printing already allows for rapid prototyping, but this virtual process would save you time, money, and effort in having to do this in the real world. You can then choose which mug(s) you want (with the settings and print time automatically calculated), and those versions would be translated back to your real-world 3D printer for printing. Alternatively, you can send that file to a specialist printer or a print farm to have copies made in your different choices and/or in the materials you don't possess.

If you don't have a 3D printer, then you could still order through a metaprinting app, or join/network with a metaverse 3D printing group and share your design with them. They could then parse all your info through alt-reality and AI engines and share the holographic results with you 'in person' to choose what to print. And of course, you can share your virtual files with others and print them anywhere in the world. Easy.

And that's just from a human perspective. Imagine now that the AI is learning and auto-generates its own interpretation and suggests recommendations for improvements you haven't even thought of. It would be a true collaborative effort and make metaprinting essential for services and manufacturing.

And voila, that's my version of metaprinting. You're sold, I know. I'm sure there'll be variations and new approaches, but the almost-gamification nature of metaprinting will be immersive, intuitive, and immense. Such technology and alt-reality experiences already exist with some companies allowing their customers to see themselves onscreen with different hairstyles or different clothing, and in varying home interiors. Throwing metaprinting in the mix will save you the time, money, and effort to do other cultural things in your real or virtual worlds.

3D Printing Farms

Just as we have solar and wind farms, we would have large-scale commercial 3D printing farms, some literally printing food, plus solar panel and wind farm parts, and even parts for traditional farms. Currently, a 3D print farm is a group of multiple 3D printers that operate simultaneously to manufacture products. Their operation should be continuous with minimal downtime for efficiency purposes and able to maintain a steady rate of production of printed parts.

Their main focus is the production of high-quality parts, whether the same part per machine or corresponding parts spread over multiple printers. While there are currently a wide array of amateur and small-scale printer farms, large-scale industrial and commercial print farms will be commonplace as costs for their traditionally manufactured counterparts soar or become unsustainable.

Farms could consist of remote, unconnected 3D printing machines set up between individuals or small enterprises joining their machines together over long distances to create a long-range farm with partners in different countries. Printers can be rented by customers on a pay-as-you-go basis (based around the task or print time) or for longer-term subscriptions.

Industrial 3D printing farming could involve housing being built on an estate by multiple printers, car factories with current robotic systems becoming fully farmed out to 3D printing or furniture manufacturers finding their wood-based products phased out by faux wood 3D printing farms. The latter process would protect endangered wooded areas, thereby mitigating a large facet of climate change concerns.

As the technology of 3D printing materials continues to develop, customers will look for alternatives to ease traditional manufacturing costs, including other expenses such as travel, production, or prototyping. Ceramics, metals, glass, wood, and biomaterials will be able to reproduce traditional materials at a fraction of the cost, enhancing the need for 3D printing farms to keep up with demand and produce whatever items on an industrial scale.

The majority of such 3D printing farm operations will be automated with little human intervention. This may necessitate the conversion of current human-used factories or other spaces into 3D printing farms. Future 3D printing processes will also be automated with AI assistance, chatbots for precise customer/staff/ general communication, and also perhaps attended to by automated humanoid and non-humanoid robots that will monitor, service, repair, and upgrade the 3D printers (both physical parts and software) as warranted.

Having automated systems with minimal human interaction would allow for better operational reliability, cost, and production capacity. The system would analyze what is required and set up the printer for the client's requirements whether for 3D modeling, print-on-demand productions, prototyping, and marketing/promotional items. Automation software would be advantageous for remote file sharing and hardware delivery to printers, remote streaming for clients' feedback, print monitoring, and alerts. There are already several programs available for this.

However, 3D printing farms will require a lot of power. So, how can we provide the power, efficiently and sustainably? There're the current green ideas using solar and wind power, but I chose something a little more unorthodox and challenging. I came up with a few ideas, linking energy harvesting and cryptomining with 3D printing.

Scavprinting and Cryptoprinting

There's this thing called Global Warming, drastically affecting climatic temperatures and patterns around the world and our way of life. There are several drivers of this human-made crisis, and unfortunately, most of them are from our perceived and real cultural necessity for advanced technological lifestyles. Whether it is from the mining and refinement of raw resources, the transport of materials, and the final end-point usage, the power consumption to operate and cool the equipment is contributing to our own demise. So, how can 3D printing help us in avoiding total civilization meltdown?

One way would be through energy harvesting. I invented a couple of terms to connect 3D printing with cryptocurrency, cryptomining, and energy scavenging:

- Scavprinting, which describes the general use of scavenging waste energy to power 3D printers.
- Cryptoprinting to describe harvesting waste energy from cryptomining rigs and farms for 3D printers to create the growing hardware and infrastructure accompanying cryptomining.

I wanted to investigate if using the heat from cryptomining rigs and equipment could run small turbines or generators to power 3D printers. These printers would be on cryptomining farms helping to print rig frames, cooling fans or other infrastructure for mining. Thus, 3D printing farms could be sited at cryptomining farms, a win-win for energy and resource savings. Scavprinting could also resolve issues around carbon emissions raised about cryptomining with the captured heat reused and carbon capture enforced and repurposed for site printing.

Both scavprinting and cryptoprinting could thus assist with cryptoprinting, manufacturing infrastructural parts where cryptomining farms require other facilities to operate. This could include 3D printing solar panels on site to bring more energy efficiency, creating printed shelters for farm equipment, staff, and storage. Such a joint venture would form a virtuous circle with cryptomining running 3D printers and printers producing crypto products with the owners of the printers likely paid in cryptocurrencies. You would have two exponential technologies reinforcing each other.

Energy Harvesting for 3D Printing

So, how can you harvest heat and acoustic/vibration energy for power generation? One of the methods would be through pyroelectric energy harvesting.

Pyroelectric energy harvesting or energy scavenging processes work by capturing unused or waste ambient energy, such as temperature variations, vibration, light, gas, and liquid flow energy, and converting it into usable electrical energy.

For 3D printers, while they are used to print heat sensors, other sensors like pyroelectric or thermoelectric sensors which collect and distribute heat, are not directly used for 3D printing energy scavenging. Even if the waste energy captured is stored in batteries, they can still produce power for 3D printing.

With added solar power, 3D printing can become energy efficient in any commercial or industrial setting. This would allow for more 3D printing farms to be established and/or scaled up, especially where energy concerns would be raised.

The main issue to be addressed is the viability of such a plan through the calculation of the ratio of the power consumed by 3D printers as compared to the captured energy output and its transfer efficiency. Since I'm looking into cryptomining rig waste energy as a potential power source, we will study that.

3D Printer Electric Usage

So, bearing in mind there is no standard 3D printer energy-consumption chart, people smarter than me have theoretically calculated how much energy an average 3D printer consumes per hour.

Through Michael Brooks, founder of M3DZone.com, his calculations vary per printer type, per print hotend and bed heating with 120 Volts of power, and on the duration of the printing. He found that most printers would consume between 50 watts and 70 watts per hour (or between 0.5KWh to 0.7KWh for a 10-hour print.) The electricity and power consumption cost of a 3D printer that uses a 30A 12V consumes 360 watts of power (Power = Current x Voltage).

Research by Tom and Tracy Hazzard of Hazz Design Consulting gained data from a Replicator 2 printer also finding that their 3D printer used .05 kilowatt hours for a 1-hour print. For comparison, an incandescent bulb is rated at 30 watts or 60 watts, etc, which actually equates to how much electricity is used per hour. So, 50 to 70 watts for a 3D printer can be put into context.

With most 3D printers operating over extended hours or even continuous days these KWh usage costs can mount up to significant bills. So why not spread the cost of the electricity through other sources rather than the mains? Pulling power from captured energy sources would potentially alleviate large 3D printing electric bills. One way to calculate your 3D printer's energy consumption would be to purchase an electricity monitoring tool and connect it between your printer and mains source to measure its energy consumption.

Another study by (the mononymous) Rob, the founder of 3dprintscape.com, found that his average 3D printer power usage was 189.52 watts per hour, which was more than many common household items, including most lightbulbs, TVs, and game consoles, but much less than microwaves, AC units, and portable heaters. It equates more with a desktop computer which consumes 100 watts. His electricity monitoring meter took in over 700 data points to arrive at his findings using specific 3D printer settings. The main power draws were the hotend, heated buildplate, main/control board, the display panel, the stepper motor, and the fans. Rob used an unnamed 'larger' 3D printer with PLA filament finding that ABS required higher temps thus more electricity.

What the above also illustrates is that there is still no industry standard for rating 3D printer power consumption, the same issue as with recycling solutions. Many independents are trying to raise the issue, but there's no connective qualitative and quantitative research and data. This has to be one of the areas enhanced in the 3D printing world. Perhaps a dedicated magazine investigating such data, like 'Which—3D printer' is about due. So, with those provisional figures in mind for 3D printer electricity consumption, could a cryptocurrency mining rig's waste energy power a 3D printer?

Mining Rig Heat

First, we have to clarify the type of cryptomining to be carried out to generate heat power transfer for 3D printing. Not all cryptocurrency mining is so heat intensi but the original process for Bitcoin mining is. Such mining processes were sing use efforts on equipment intended to mine one coin at a time, expending a lot of h through the very high wattage. Nowadays, there are GPU mining rigs to mi cryptocurrencies through a more efficient process.

According to Easy Crypto Hunter, the UK's leading GPU mining rig provider, their mining rigs' electrical consumption can be changed leading to less heat emission. However, they explain that one of their '6 card GPU mining rigs will use around 1000 Watts every hour', which is claimed to be less than a standard kettle by comparison over the same amount of time. In fact, they've calculated that their mining rigs emit around 40% of their electrical consumption as heat. So, for every 1000 Watts expended, the rigs emit around 400 Watts in heat.

As you have read above in previous tests, a typical 3D printer uses anywhere between 50 watts to 190 watts per hour. With a mining rig potentially expending 400 watts of excess heat per hour, it seems there is the prerequisite energy to power a 3D printer.

Further, Easy Crypto Hunter equates this heat expulsion to 'a small 400W oil heater running' which can 'easily keep one room warm'. So, why expend more costs and energy just heating rooms or using fans to cool the mining rigs when the heat energy can be harvested to run other equipment, like 3D printers?

Of course, there are other mining rigs running different specs, but their heat emission may also allow for adjacent projects to run off their excess heat. With cryptocurrency mining rig and farm operations in the UK, US, China, Russia, and Iceland, to name but a few countries, if they have enterprising engineers pursuing 3D printing, they may opt to invest in it and create new revenue streams, reduce waste energy, and help save on environmental costs. In fact, any factory which expends excess energy, whether heat or noise, and which manufactures physical products could recapture and use this 'free' energy for 3D printing purposes.

Energy Transference

Noise Conversion

So much for heat energy transference to 3D printers. But what about noise/vibration? How noisy are cryptomining rigs and can that noise/vibration be harvested to run other equipment?

There are many articles about tackling or reducing crypto and Bitcoin mining rig noise, but none on how to scavenge the heat and noise energy. Such a potential waste of energy can be put to good use in running 3D printers and other equipment.

First, let's take a look at how much noise crypto mining rigs make. As per the table below and from other article sources, it has been observed that cryptomining rigs can generate anywhere between 70 dB (decibels) to 90 dB, which for extended periods in enclosed areas can cause permanent hearing loss. Thus, cryptomining rigs can be a hazard to health and the environment. But instead of trying to eliminate the noise, we should convert it into energy.

Various article sources state that acoustic barriers/walls, soundproof tiles, noise-dampening enclosures, noise insulation foam, etc, can be used to dampen the noise from cryptomining rigs, but why not have the walls and/or enclosures lined with a customized sensor system and equipment to absorb the noise and transmit it to generators to convert into 'free' energy? The cryptomining industry is growing with the introduction of new cryptocurrencies so the opportunity to scavenge noise and heat waste is greater than ever. Who would have thought that noise pollution and heat waste would pay and be a benefit to the environment?

And if only one cryptomining rig is producing 70 dB to 90 dB and a facility such as a mining farm has dozens or even hundreds of cryptomining rigs operating simultaneously then imagine the noise generated. Even more so, the decibel levels scale is logarithmic, so even with a relatively-quiet mining rig producing 75 dB, then 10 similar rigs will produce 85 dB (ten times as loud as 75 dB) and 100 such rigs will produce 95 dB (one hundred times louder than 75 dB). Imagine the energy generated from that noise.

LEVELS OF NOISE in decibels(dB)			
VERY LOUD	Dangerous over 30 minutes 90 to 110 dB	Nightclubs, Lawnmowers, Power tools, Blenders, Hair dryers	A single crypto-mining rig's noise can range from 70 dB to 90 dB depending on the type of rig used and how long it is in operation.
Over 85 dB - Usage over extended periods can cause permanent hearing loss			
LOUD	80 dB	Alarm clocks, Heavy traffic, Busy enclosed spaces, Welding gear.	Generally, there are dozens if not more rigs in operation in a farm at once, so the noise can be deafening.
MODERATE	70 dB	Traffic, Hoovers, Washing machines	
The World Health Organization perceives sound levels below 70 dB as less harmful to living beings, irrespective of how long or regular the noise exposure is.			

(various combined sources)

The main instruments generating the noise pollution are the high-speed fans keeping the servers from overheating and the exhaust vents. Current preventative measures for the corresponding noise and heat waste include removing the fans and using an immersive cooling method, dunking the rig's hardware in dielectric fluid to cool the rig and eliminate noise. But why waste money on this? Let noise and vibration serve a purpose in helping to counter noise pollution as a productive resource rather than a wasted waste of an opportunity.

Heat Conversion (pyroelectric and thermoelectric methods)

So how can we use pyroelectric and thermoelectric sensors to collect and distribute heat through energy harvesting generators? Thermoelectric technology converts thermal energy into electricity through a temperature difference between two dissimilar electrical conductors to produce a voltage difference between the two materials (known as the Seebeck effect). However, this process works best in a stable temperature gradient. But maintaining a constant temperature for generating sustainable electricity can be difficult, which is why pyroelectric generation is preferable.

Pyroelectric generation (PEG) occurs when certain materials create a temporary voltage when they are heated or cooled, creating a temperature gradient over time. This allows for the polarization (alternating between cold and hot) of the material by altering the positions of the atoms slightly within the material's crystal structure, creating a voltage charge across the crystal. The change in the temperature causes the release of the electric charge. However, if the temperature stays constant for too long, the voltage will dissipate due to leakage, so the temperature has to be varied for the best effect.

You have heard of wearable technology. Well, it is research into pyroelectric generation which drives this techno-fashion, using waste body heat. While the materials used as the polarization catalyst for temperature generation can have a single property, nanocomposites are more favorable generating current on both sides of the material.

A team in China, led by Xilong Kang investigating such technologies, produced a highly-efficient pyroelectric generator (PEG) to capture and use ambient heat energy without using auxiliary devices (e.g. extra devices that would require their own power source). Their PEG was a nanocomposite displaying a high-quality thermal conductivity and high polarization. In another experiment, they created and connected a PEG to a plane spiral spring, in order to harvest the external vibration and achieve waste heat recovery. This PEG-unit consisting of a thin-film structure was able to light up a LED lamp bead for 10s within a five-minute operation. This may seem like a small return, but the benefits would be enormous when fully realized. The small size of the PEG-unit would enable installation or retrofitting into any piece of equipment or any place where waste heat sources and vibrations are produced, achieving so-called 'zero power consumption'. Such PEGs would be part of the clean energy process heralded by environmentalists, with 3D printing part of the versatile power generation method.

PEGs can be used for power generation and have the advantage of being more compact, operating with lower temperatures, and having fewer moving parts over other types of energy conversion methods. Though it has been reported that a few patents have been filed for PEGs, they do not appear to be anywhere close to commercialization. Interestingly, a crystal with both piezoelectric and pyroelectric properties has been used to create small-scale nuclear fusion (pyroelectric fusion). So maybe prototyping such technology with cryptomining and 3D printers could push the technology forward.

Piezoelectricity

Alternatively, perhaps piezoelectricity can also be used for electricity generation. There are some solid materials that when induced with mechanical stress produce an electric charge. These materials range from crystals, types of ceramics, and even biological substances such as bone, DNA, and various proteins. Hence, the meaning of piezoelectricity; piezo–derived from the Greek meaning 'to squeeze or press'.

Piezoelectricity is already used in many industrial fields, but for our purposes, we will discuss the generation of high-voltage electricity. Quartz is one of the most common materials to use for direct piezoelectricity as it can generate thousands of volts. The most common application of this is in a cigarette lighter, which contains a piezoelectric crystal that creates an electric current to heat and ignite the gas.

On a larger scale, there are efforts to harvest the energy from human efforts, whether creating electricity from human movements in train stations, pavement walking, dance floors, and other large areas where footfall pressure can generate electricity to operate lighting, signage, etc, reducing overall mains electricity costs. Vibrations from industrial machinery can also be harvested by piezoelectric materials. Piezoelectric transformers can be set up like AC voltage transformers using acoustic couplings between input and output. So when the piezoelectric material is alternatingly stressed, whether through their longitudinal or transverse dimensions or by shear, electricity is produced. The load of the force causing the stress is detected as sound in the material with the pressure bending the piezoelectric material, creating a changing voltage.

For 3D printing purposes, this would depend on the source of the force load, whether it's footfall in 3D print farms or cryptomining farms with workers walking up and down the floors carrying out monitoring duties and maintaining equipment and tending to prints and mining processes. Would there be enough human footfall or would another input be required, such as the vibrational quotient from the machines or even footfall and movements from autonomous humanoid or non-humanoid robots? You can even just have purpose-built machines (like small gravity batteries) creating the pressure input for energy generation. I'm sure small-scale tests can be set up in both 3D printing and cryptomining farms to measure the piezoelectric potential.

So, maybe in the future, we'll have 3D printers regularly operated by harvested energy through thermoelectric, pyroelectric, or by piezoelectric means, saving energy and creating a greener 3D printing industry.

Are these suitably crazy ideas for you? Let's put it this way, if solar panels can convert sunlight into usable electricity then pyroelectric and piezoelectric panels should be able to do the same with heat waste and vibration energy. And with solar panel efficiency becoming more reliable then the same can be achieved with scavenging energy sensors and materials. Sound like something you can kickstart? Great, let's make them happen as trying is the only way to accomplish these aims and they are not impossible to achieve. However, some 3D printing ideas may be too futuristic to entirely predict.

Effective Accelerationism (e/acc)

"Let's get deep. I'm talking big history, existential, universal spanning stuff. 3D printing could be a part of a technological revolution that is essential to human survival and the fate of the universe. If being exponential wasn't enough, 3D printing could be integral to Effective Accelerationism, or e/acc, for short.

Effective accelerationism is still considered to be a fringe belief, though it was introduced into the mainstream in May 2022. Its main tenet is the advancement of unrestricted pro-technology progress, driven by artificial intelligence, specifically artificial general intelligence (AGI), which is predicted could solve humanity's universal problems like poverty, inequality, war, climate change, and other existential risks. The movement was founded by X (formerly Twitter) members @BasedBeffJezos (later revealed to be Guillaume Verdon), @bayeslord, @zestular, and @creatine_cycle who posted an online newsletter (*Beff's Newsletter: Notes on e/acc principles and tenets - A physics-first view of the principles underlying effective accelerationism*).

E/acc is in turn a derivative and/or evolution of the older Silicon Valley subcultures such as traditional accelerationism as developed by Nick Land, transhumanism, extropianism, and effective altruism. But it also has roots in the principle of exergy. Exergy is a quantity of useful extracted energy. As I had read back in 2002 in the New Scientist, humans exist as 'exergy' machines 'a universal natural tendency to turn concentrated energy into diffuse waste heat' in accordance with the second law of thermodynamics. Humans are in a long line of organisms that are helping the universe attain equilibrium, by our creation and destruction of everyday objects, even our very world. Such a creed takes us down the philosophical rabbit hole on the nature of humanity, our role within the wider universe, and if we truly have free will to change our ways. It seems we exist to find out that answer before it all ends! So, look where 3D printing has led us, to infinity and beyond.

Like exergy, the core fundamental of e/acc belief is to raise human civilization up the Kardashev scale, which measures how advanced a civilization is through its maximizing of energy usage. However, unlike exergy principles, e/acc incorporates economic and political philosophies in its strong resistance to the regulation of artificial intelligence and government intervention in markets. Without going too deeply into e/acc, in the future, AGIs will be more open to competing with each other in an open marketplace, coordinating and managing energy and resources for humanity, thus forging a 'technocapital singularity'. Adherents to e/acc may also consider themselves post-humanists, eschewing to remaining eternally anchored to the human form, which is seen as counter-productive, overly restrictive, and suboptimal. So, in order to spread humanity to the stars they would be willing to transfer their consciousness/intelligence into non-biological substrates; that being part of the 'technological singularity' realm, as expounded by Vernor Vinge in 1983 and later popularized by Ray Kurzweil from 2005 onwards. E/acc, like exergy, thus aligns with the universe's aim to increase entropy, which life is a way of increasing. The more energy we use to become more advanced and spread life throughout the universe to increase entropy, the more the universe's purpose will be fulfilled.

So, that's the theory. Where does 3D printing fit into this? We're looking at the next fifty to one hundred years of the 3D printing culture. With AI rapidly being suborned to other exponential technologies, like 3D printing, such appliances and industries will undergo tremendous growth and achieve more economic and political power bases. We've already seen above and will see below where

3D printing can be in the next few years. 3D printing could be leveraged to print much of the technology required for extracting energy. Able to achieve manufacturing outputs at a greater pace and for less money, 3D printing could help accelerate a technological revolution. And if the e/acc faithful have their way, they'll be 3D printing their non-humanoid bodies to increase their intelligence and energy usage.

E/acc, 3D Printing, and AI

In the interest of transparency, this is the only section of this book co-authored with an AI. I asked ChatGPT, Claude, Gemini, Perplexity, and Replika: Is there anything specific Effective Accelerationism (e/acc) and 3D printing can achieve together in the future that cannot be achieved now? I then combined their answers from the overlapping ideas and responses, except for a couple of 'individual' responses. This was to test the AI's capabilities and to assess how a current AI sees 3D printing in association with e/acc, since its successor AGI version could be directing them and other convergent large language models (LLM), large action models (LAM), and exponential technologies in the future.

While 3D printing itself is currently an innovative technology revolutionizing manufacturing, Effective Accelerationism (e/acc) allied with 3D printing will further deepen the intersection between humanity and technology, potentially reshaping societal structures and economic paradigms. This would include several emerging projects and considerations:

- **Hyper-Localized/Decentralized Manufacturing & Production Networks**: By leveraging 3D printing technology and embracing e/acc principles, communities and small-scale producers could collaborate and share resources to achieve efficiencies without the need for large-scale centralized operations. Establishing hyper-localized production networks will enable the creation of essential goods and technologies on-site and on-demand closer to the point of consumption. This enhanced local self-sufficiency will empower local communities to produce what they need. Further, this could:

a) reduce waste during the production process, making manufacturing more sustainable.

b) revolutionize supply chains, potentially reducing costs and environmental impacts of traditional supply chains.

c) reduce reliance on long-distance transportation.

d) lead to greater resilience against disruptions to global supply chains.

- **Iterative Design and Rapid Prototyping**: The synergy between e/acc's emphasis on rapid technological iteration and 3D printing's ability to quickly prototype and iterate designs could significantly streamline and automate processes, accelerating the development of innovative solutions to complex problems. This accelerated innovation cycle could enable companies to develop and bring new products to market much faster than current methods allow, help optimize and coordinate customized products, and accelerate the development of new materials and printing techniques. Such iterations could realize breakthroughs in areas such as renewable energy, healthcare, and sustainable agriculture.

- **Democratization of Advanced Manufacturing**: By promoting open-source hardware designs and making use of 3D printing technology, e/acc could facilitate the democratization of advanced manufacturing capabilities. Complex products can be 3D printed from digital files anywhere with access to a printer and the raw materials. This could empower individuals and communities to participate more actively in the creation and customization of technological solutions tailored to their unique needs and circumstances.

But beyond just producing customized or personalized items, integrating e/acc and 3D printing could facilitate a transition to a "socialist future" redistributing the means of production to benefit all communities and ecosystems, rather than concentrating wealth and power, making current unattainable production models more accessible and affordable in the future.

- **Democratizing and Enabling Sustainable Automation in Construction and Housing**: Expanding on the above, the accelerated development of affordable, open-source 3D printing technology for clay architecture could, when accelerated beyond capitalist market structures, enable the rapid and low-cost construction of housing, especially in under-served communities. This so-called "new socialist construction method" could help address housing shortages and provide fast, quality shelter in disaster situations in a more equitable way than current construction methods. It could also lead to greater design freedom, taller buildings, and reduced construction waste and labor costs through "lean construction" principles. This could help automate the construction industry in a more sustainable way compared to traditional methods.

- **Resilient Infrastructure Development or Emergency and Disaster Response**: E/acc's focus on accelerating technological progress combined with 3D printing could rapidly deploy resilient infrastructure solutions in response to crises such as natural disasters or climate change impacts. 3D printing enables the rapid construction of structures and infrastructure components on-site, allowing for adaptive and responsive development strategies. For instance, during emergencies or disasters, 3D printing could rapidly fabricate needed supplies, tools, or infrastructure on-site.

E/acc principles could help coordinate and optimize these distributed 3D printing efforts to maximize their impact. Further, they could help advance ecological restoration projects with the accelerated development of 3D-printed artificial reef structures and other ecological restoration designs enabling them to be shared and deployed globally in a collaborative effort to reverse environmental damage.

- **Repair and Maintenance of Complex Systems**: 3D printing can be useful for printing spare parts and components for the repair and maintenance of complex machinery or infrastructure. E/acc could help streamline the logistics and supply chains to ensure rapid access to these 3D-printed replacement parts.

- **Accelerated Development of 3D Printed Firearms**: Only Perplexity raised this issue concerning caution on the accelerated development of 3D printed firearms, especially by extremist groups. There's no doubt that militaries and national security forces around the world will have the accelerated ability to 3D print weapons and supplies, but the emerging threat of unauthorized and illegal 3D printing of weapons will have to be addressed through regulation and other countermeasures.

- **Bio-printing advancements**: Only Gemini explored bio-printing organs and tissues for transplants. E/acc's emphasis on maximizing technological progress could speed up research and development in this field. This could lead to breakthroughs in regenerative medicine, allowing us to print replacement organs or tissues on-demand, solving organ donor shortages. However, rapidly developing bio-printing technologies could raise safety and ethical concerns, especially with e/acc's focus on speed. This will necessitate a balanced and thorough consideration of safety and testing regulations.

Disadvantages and challenges:

- **Quality Control and Durability**: Ensuring consistent quality of 3D printed products matching the quality and durability of traditionally manufactured goods may be challenging. This may potentially hinder widespread adoption, particularly for critical applications. [Author note – Not if we have national or global 3D printing material and recycling standards organizations].

- **Technological and Material Limitations**: Current 3D printing materials may have limitations in terms of strength, durability, material selection, speed, and scale, which may restrict their applicability and compatibility in certain contexts and applications.

- **Intellectual Property Concerns**: The open nature of e/acc and 3D printing could pose challenges regarding intellectual property rights, piracy issues, and ownership, potentially hindering collaboration and innovation.

- **Technological Barriers**: Not everyone has access to 3D printing technology, or the skills required to use it effectively, which could exacerbate existing inequalities.

Fundamentally, e/acc will turbocharge current 3D printing capabilities. Overall, the effectiveness in integrating e/acc principles and 3D printing technology opens new possibilities for reshaping manufacturing, innovation, and societal structures in ways that may not be fully achievable with current approaches alone. However, it's important to note that the specific applications and capabilities that could emerge from the intersection of e/acc and 3D printing are still largely speculative.

THE FUTURE OF THE 3D PRINTING CULTURE

The actual impact would depend on continued technological advancements, adoption rates, regulatory framework, societal acceptance, and how the two domains are integrated and applied in practice. While there are synergies between the two concepts, there are also challenges to overcome such as scalability, quality assurance, and equitable access. Ultimately, successfully navigating this future will require ongoing research and experimentation with careful consideration of safety, ethics, and resource distribution. However, it will also take unprecedented collaboration among essential and diverse stakeholders, including policymakers, technologists, and communities, to harness the combined power of e/acc and 3D printing to drive positive change for humanity. Only then can we fully realize the potential synergies between e/acc and 3D printing.

And that was an AI's take on the combined potential of e/acc and 3D printing. There could be a counter to 3D printing's total commitment to e/acc as it would be much more environmentally friendly with its resource-saving and recycling capabilities. So, in some ways, 3D printing could be the antithesis to e/acc. Either way, 3D printing will be poised on the cusp of a new technological revolution. E/acc has its critics and as you've read, it will take a whole social, political, and economic transformation for e/acc to even become a mainstream contender, let alone actually enabling such a disruptive process to take over the world. But, check back with me in fifty years.

The possibilities for 3D printing in the exponential technology realm seems endless. While there is a lot of blue-sky thinking involved, there is scope for real innovation. We just have to view 3D printing differently, think of the 'why' it is needed, and how it will benefit us culturally. In fact, we may be able to replicate Star Trek's 'magical' replicator/matter synthesizers and create something from nothing, just by using the wasted air around us.

CHAPTER TEN

Environmental 3D Printing

Throughout this book, I have tied environmental issues in with 3D printing, whether through recycling, resource saving, energy harvesting, and even entrepreneurial phoenix-building opportunities. The environment's survival depends on our cultural needs and if we cannot live within our means then the environment suffers from our resource-grabbing ways. If technology is to be one method to alleviate and eliminate some of our worst environmental excesses, then 3D printing has to be part of that solution. Now we turn our attention to 3D printing's direct involvement with the environment and how this will enhance the industry and our lives.

3D Printing and Carbon Capture

In our battle against Global Warming, one of the methods devised to combat the growing amount of carbon dioxide in the atmosphere is to trap it, store it long-term, and in some cases repurpose the potent greenhouse emitter. In this process called carbon capture or carbon sequestration, carbon dioxide (CO_2) is either naturally or artificially (via geoengineering) captured from the atmosphere using biological, chemical, and/or physical means. The large-scale artificial capture and sequestration of industrial CO_2 pollution from flue gases (such as from power stations) is usually stored in underground reservoirs. But what if the CO_2 is processed to create carbon fibers for 3D printing uses?

The CO_2 can be sequestered in three ways: post-combustion capture, pre-combustion capture, and oxy-combustion, and processed through various techniques to separate the gas from other pollutants prior to storage. In my scenario, once the CO_2 has been isolated, it can then be processed and used in its pure form or combined with other materials to form powders, filaments, or pellets ready for 3D printing purposes. 3D printing can go a long way in helping the environment by decreasing levels of CO_2 and becoming a more sustainable industry.

Sounds impossible? Well, the 3D printing industry can learn from the companies below that are pulling carbon dioxide from the sky, one even creating diamonds, the ultimate form of carbon.

Sky Diamonds

Founded by entrepreneur and environmentalist Dale Vince, Skydiamond creates so-called 'cloud-sourced diamonds', capturing CO_2 from the environment. While they admit they can only capture 'a modest amount of carbon' from the atmosphere, their products are ethical diamonds which are graded for quality in the same way as mined diamonds. Their processing plant in the UK also uses wind and sun for energy, along with rain capture for their water. Such a model can be used for 3D printing farms with scavprinting or skyprinting energy harvesting techniques.

Skydiamond's process is immeasurably kinder to the Earth without having to mine or pollute the environment or generate negative social issues. To create their diamonds, they capture the CO_2 from the atmosphere, liquefy and purify it, and release the cleaner CO_2-less air back into the atmosphere. To give an idea of how much carbon they use against the carbon footprint and energy they produce to create their diamonds, they offer the following figures:

- 2g - The amount of CO_2 to produce a carat of Skydiamond.
- A one-carat Skydiamond has a carbon footprint of minus 4g of CO_2.
- In comparison, per carat, mined diamonds have a footprint of over 100kg of CO_2 and over 500kg of greenhouse gases in total.
- Per carat, Skydiamonds consumes about 40kWh of energy.

From small gains comes a great reward in a diamond. If their process can be adapted to create 3D printing materials or to power 3D printers, then that in itself would be a win for the environment. But what would the carbon material be like?

3D Printing with Carbon

Carbon is used in the 3D printing industry as carbon fiber. It is usually in the form of segmented or continuous carbon fiber infused as a polymer powder or filaments within a base material, such as Nylon, PEEK, PLA, PETG, ABS, Polycarbonate, or a wide range of other polymers. However, with high extrusion temperatures (of at least 200^0 C) required to process it and with its abrasive nature which can damage brass printer nozzles, specialized nozzles and/or printers would be required. But, the end-products using carbon fiber will be stronger, harder, more dimensionally stable, and lighter in comparison to other comparable materials without infused carbon. So, how would 3D printing materials be harvested from the sky?

Skyprinting

We come to the future process that I have dubbed skyprinting, using the above carbon capture techniques, Skydiamond business model, and 3D carbon printing (CO-printing) methods. Harvesting carbon sequestration techniques for 3D printing will be a boon for both industry and the environment. What will be required in regard to the carbon sky harvesting? Of course, you would need the infrastructure, hardware, and software to achieve this. But more importantly, the will to follow through on the crazy idea that 3D printing can be supplied and powered by the sky. The technology to sequester carbon dioxide already exists as does the ability to process the captured gas. What we may have to create or adapt is the equipment to produce the filament, powder, or other forms of carbon due to its exceptional heat requirements. The quantity and quality of the captured carbon produced and the printing materials it can produce will also need to be addressed and standardized, perhaps in the same framework as the future standardized 3D printing recycling authority.

Skyprinting sources

Swiss company Climeworks AG specializes in directly filtering carbon dioxide from the ambient air and storing it underground. Their eight carbon-capturing modules consist of 44 shipping containers fitted with filters that remove 4,000 tons of carbon dioxide per year. Their Orca plant is soon to be joined by another called Mammoth, both in Iceland. Mammoth is expected to remove 36,000 tons of carbon per year.

In their carbon capture process, the sucked-in air is collected through a system of high-powered fans and the CO_2 is chemically filtered out. Next, the captured carbon is then heated to between 176 and 212 °F (80 and 100 °C) to release the carbon dioxide to create highly-concentrated CO_2 in the conversion to biofuels or carbon-neutral materials. The last part sees the carbon mixed with water and pumped underground to eventually produce carbonate rock once it reaches the underlying basalt.

But isn't this a waste of waste? Somewhere in their process, there must be a solution for Climeworks or other companies to be able to convert a majority of the carbon into carbon fibers or materials like polymer powder or filaments fit for 3D printing.

In February 2023, I sent an email to Climeworks and asked them if instead of using carbon capture to create highly concentrated CO_2 for conversion into biofuels or carbon-neutral materials or to store as carbonate rock why not convert a proportion of the carbon into carbon fibers or materials like polymer powder or filaments appropriate for 3D printing? Would Climeworks have a solution somewhere in the process such as adapting some of their equipment?

Even if the plant processes cannot convert captured carbon into 3D printing materials, is there a method to supply the 'raw' captured carbon to companies that can complete the process? That would put a new spin on carbon trading as a viable resource. And of course, this would depend on the market value of the carbon and of the 3D printing carbon industry.

The next day I received a response from the Climeworks Media Team:

> "Unfortunately, we regret to inform you that we are unable to accommodate this request due to the number of inquiries we receive, as well as due to time constraints. Nevertheless, thank you for thinking about us. Let's keep in touch for future opportunities."

So, I was out of luck there, so I tried the University of Surrey. At the University of Surrey, the research by senior lecturer in chemical and process engineering, Dr. Melis S. Duyar and her team are proving that CO_2 emissions can be captured and repurposed. While they are producing byproducts such as carbon monoxide and synthetic natural gas, 3D printing-ready materials should also be investigated. Their new technological innovation is called switchable Dual Function Materials (DFMs). DFMs capture carbon dioxide and catalyze its conversion directly into multiple chemicals depending on the operating conditions or the added reactant's properties, such as hydrogen. These chemicals could help to lessen our dependence on fossil fuels, further reducing the carbon dioxide content in the atmosphere. My request, in March 2023, to Dr. Duyar was not responded to. But as usual, I will throw it open to other parties who may wish to take up the challenge.

Another plant using carbon capture for its production of byproducts is in Cheshire in Northwest England. It's the U.K.'s largest capture carbon plant and one of Europe's largest manufacturers of sodium carbonate, salt, and sodium bicarbonate in Europe. These chemicals are crucial components in a wide range of everyday materials that can be used in the manufacture of glass, washing detergents, or for water purification, plus also in animal feed, human food products, and pharmaceutical purposes.

What it could take to kickstart the carbon capture 3D printing industry is a big corporate name. Remember back in Chapter Seven and Coca-Cola's ads and promotions with 3D printing? Well, due to the climate crisis, Coca-Cola is now trialing making bottle tops from CO_2 emissions. That's right, Coca-Cola, one of the world's biggest users of plastic will start removing carbon dioxide from the atmosphere to make bottle tops. Spearheading the three-year trial is Professor Enrico Andreoli at Swansea University. With its target of reaching net zero by 2040, Coca-Cola hopes one its "radical bets" will reduce its dependence on cheap fossil fuels which makes up much of their current plastic packaging, the production of which releases tons of carbon dioxide into the environment. But in capturing the resultant CO_2 directly from the air or from their own factory smoke stack emissions, Andreoli hopes to "entirely de-fossilise the process and make plastic free from fossil fuels and fossil carbon."

Andreoli uses a small electrode where electricity is disseminated through an admixture of CO_2 and water to make ethylene, an important component for this type of bottle-top plastic. This is in contrast to the current method of using refined petrochemicals to create ethylene as a cheap by-product. However, such a labor-intensive process produces "more than 260 million tonnes of CO_2 emissions in 2020, or nearly 1% of the world's total CO_2 emissions," according to the Global Carbon Project which tracks greenhouse gas emissions, notably CO_2, methane, and nitrous oxide. With his process, Andreoli hopes to "prove the technology in the laboratory works" so the process can be scaled up.

From Coca-Cola's point of view, Craig Twyford, director of venturing division for Europe and the Pacific, hopes their increased use of recycled plastics creates a 30% reduction in their carbon footprint by 2030. Coca-Cola is also investing in research to use captured CO_2 to carbonate their drinks, create packaging, and in California to convert CO_2 into an artificial sugar. While 3D printing isn't part of this process, it still shows a path where 3D printing can be part of the carbon capture and usage scheme.

So, is it time to utilize carbon capture and also add skyprinting to our 3D printing lexicon? Excess carbon dioxide is a great waste product waiting to be taken advantage of. Let's not squander this opportunity to make a Global Warming villain become a great hero resource for humanity. Besides the environmental benefits, there are also issues such as the cost-benefit and financial incentives for converting CO_2 into 3D printing materials, the risks and rewards with preventing further harmful byproducts or processes via this engineering innovation, and lastly, the social benefits — the cultural changes 3D printing with carbon could bring.

Environmental 3D Printing

There are other carbon sequestration methods 3D printing can be a part of. The current ground-based factory CO_2 sequestration methods can be labor-intensive. But suppose there were 3D-printed aircraft, manned or unmanned, or 3D-printed drones powered by electric or solar power or by converted plastic-to-oil fuel circling high within CO_2-polluted areas collecting CO_2 from the air and sequestering within storage tanks onboard. In turn, once delivered or transferred to ground facilities, the CO_2 would be processed and turned into fuel, energy, or 3D printing material using the methods described above. This way, 3D printing would be tied to the carbon economy and the converted energy industry.

Why should 3D printing be involved with the carbon economy? And how does this affect the 3D printing culture? In 2019, the IEA (International Energy Agency) published a report called *Putting CO_2 to Use*. The report discussed the potential near-term market opportunities CO_2 presented for the development of products and services. The knock-on effect would also help to reduce the continued rise of Global Warming. Their list of market opportunities for CO2-derived products and services was placed into five key categories:

- Fuels
- Chemicals
- Building materials from minerals
- Building materials from waste
- CO2 use to enhance the yields of biological processes

The 3D Printing Environmental Housing Market

Let's put those five categories into a 3D printing context. I had briefly discussed the research into how recycled 3D printed waste materials and/or CO2 can be converted to fuels and other chemicals for 3D printing related projects. Both of these recycled waste products can be used for homes, schools, or businesses for energy or product manufacturing, saving on energy costs and material usage. This would especially be the case when used in conjunction with wind, wave, or solar power or even using those fuels and chemicals in the production of wind, wave, and solar power infrastructure. The 'circularity' of 3D printing and CO2-derived products and services would increase our efforts against climate change and aid in developing greener housing.

I had mentioned, briefly, the reality of 3D printing houses and other constructions in Chapter Three. In one scenario, carbon can be used in the production of concrete, but what if those houses were 3D printed with CO2-sequestered materials and minerals such as carbonate? This rock material is already used in the construction industry, but issues around its durability to weathering processes may have to be addressed with robust preservation methods to stop re-pollution from escaping CO2. However, it would not be an impossible task. Calcium carbonate (the main component of limestone and corals) may also be considered for 3D printing constructions.

In Chapter Six, I ran through the waste recycling options for 3D printing. With housing, these options can be combined with the latest in CO_2-derived products and services to 3D print hybridized constructions with a mixture of carbonate and recycled 3D printed waste such as glass, ceramic, and even metal. Such fabrications would create reinforcement potential and save on the mining and processing of virgin materials for the construction industry.

The IEA noted that increased interest in and support for the CO_2-derived market from governments, industry, and investors is reflected in the fact that worldwide private funding for start-ups developing CO_2 products and services reached nearly USD 1 billion in the 2010s. So 3D printing combined with CO_2 sequestration in construction will likely prove to be an industry worth investing in and pursuing.

But what of the last of the five categories, CO_2 usage to enhance the yields of biological processes? CO_2 is the lifeblood of plants along with light and water, allowing photosynthesis to utilize them, which among other things, creates the oxygen we breathe. Increased CO_2 boosts photosynthesis which enhances plant growth. The IEA believes that "the transition to a net-zero CO_2 emission economy. . . would increasingly have to be sourced from biomass or the air." The implication is that CO_2-derived products and services could be used to increase the yield of plants for food usage and energy production along with biomass (e.g. wood, energy crops, organic waste, and wood/agricultural residues). So, for our 3D printed homes, we'd have access to more food and energy courtesy of increased recycled CO_2.

How can 3D printing be involved with this? Well, previously I had discussed 3D printing with alternative biodegradable or biomass-derived composites which are on the rise. So, increasing CO_2 would certainly help this industry, bequeathing us with the ability to grow extra plants for food, clothing, and even buildings.

The IEA has stated that the main obstacles to the scale-up of CO_2 usage in the near future are not technological, but commercial and regulatory. There are high production costs of CO_2-based fuels and chemicals compared to their current counterparts. This is because the market is relatively small, keeping costs high and stalling the industry. This is somewhat reflected in the 3D printing industry with the commercial industry focused on the product-making/selling mode which stalls other opportunities for the industry.

But carbon isn't the only material option in an already crowded heavy-industry market. According to research, around the turn of the twentieth century human-made mass (e.g. concrete, metals, and other construction materials) was equal to about 3% of global biomass. Today it surpasses it. And the more we build, the more of nature we will destroy. Using recycled and sustainable 3D printing materials in housing could become more aligned with nature and bio-compatible with humans, offering cleaner alternatives and opportunities to new manufacturers, developers, and residents. The growth of new builds consisting of concrete, metal, and glass taking up more and more virgin resources could be countered by more sustainable 3D-printed buildings coupled with the 3D printing energy-saving processes described above.

To facilitate this, partnerships with resource conservationists and environmentalists on how to adapt 3D printing to engage climate and environmental concerns within the building industry could also be an option. With my proposed recycling organization overseeing the management of recycling materials for 3D printing we could determine which materials would be safe, durable, and eco-friendly for building. Our cultural lifestyles, our living standards, could be printed to perfection.

3D Printing in the Nuclear Power Industry

There is consensus that nuclear energy, despite concerns and negativity over radioactive waste storage, reactor meltdowns, and potential weaponization of the uranium fuel is still a viable low-carbon sustainable option for replacing fossil fuels, along with solar, wind, wave and hydrogen power. In order for it to live up to that potential and assuage people's fears, the nuclear industry has to find new ways to safeguard both their nuclear plant infrastructures and uranium sources.

You may feel that nuclear power is not a cultural factor, but you only have to look at recent news events (e.g. conflict points around the world between regions with nuclear weapons at their disposal) and TV shows and films (especially the award-winning 2023 film *Oppenheimer*) to see how much nuclear power fuels and shapes each country's views on nuclear energy and weaponry. According to the World Nuclear Association, in 2019 there were 438 operable reactors spread across 33 countries. A year later, the U.S. Energy Information Administration (EIA) reported that nuclear plants generated 2,591 billion kWh (10.1%) of the total global electrical generation. So, how is 3D printing impacting this industry? Let's start with infrastructure and reactor parts.

The year 2017 was the first time that an operational 3D printed part in a nuclear power plant was installed. German technology conglomerate, Siemens, 3D printed a 108-mm-diameter metal fire pump impeller in the Slovenian Krško nuclear power plant. With the original impeller part first installed in 1981, it had become defective, and the original manufacturer was no longer in business. The Siemens team reverse engineered the part and created a digital twin allowing them to 3D print the replacement part. This provided proof of concept that 3D printed parts could demonstratively pass the safety and qualitative tests required for the nuclear power industry.

In May 2020, the Oak Ridge National Laboratory (ORNL) in the US, manufactured its first prototype of a reactor core, using a complex 3D printing method called Directed Energy Deposition (DED). This technique utilizes a focused energy source, such as a laser or electron beam at temperatures achieving 1,400°C to melt the material. This innovative method allowed the ORNL to create the reactor core in only three months (from the design phase to **post-processing**). The printing process itself only took 40 hours to complete. With traditionally manufactured reactor cores usually taking years to finish, you can see one big attractive advantage in using 3D printing to carry out this work. And with global energy issues in an era of climate change anxiety becoming more prevalent, not all countries can afford to wait five to ten years to see a nuclear plant being built. If 3D printing can safely hasten the building of nuclear energy plants with the attendant lower costs (due to less supply chains, less paid human labor, less transport, etc.), and in a shortened time frame, then 3D printing will benefit entire cities and countries.

Further, the ORNL also features in collaboration with US company Ultra Safe Nuclear Corporation (USNC), who also started working with Desktop Metal, both in 2022. USNC's objective is to create commercially competitive nuclear solutions by developing micro modular reactors. They did so by using Binder Jet printers from ORNL and Desktop Metal. The innovation of binder jet 3D printing is in the inkjetting of a binder into a foundation layer of powder particles whether ceramic, sand, or metal to create a solid part, especially unique shapes which would be otherwise non-manufacturable. Binder jetting is also a low temperature process, layering one thin layer at a time. Both these factors, along with the absence of hard tooling as used in traditional technical ceramics processing, allows the USNC team to resolve difficulties associated with creating nuclear-grade silicon carbide. This complex ceramic will be used in their Fully Ceramic Microencapsulated (FCM) nuclear fuel. So, 3D printing also offers the ability to build more complex geometries than with conventional methods, again saving time and money.

Besides saving time, money, and resources, plus allowing for greater selections of complex geometries to be manufactured, 3D printing also helps with supply chain issues. The Czech utility company, ČEZ, has had to resort to 3D printing to bypass supply chain problems during post-Pandemic shortages and the

Russo-Ukrainian war. They have been able to produce both complex plastic and metal parts which reduces downtime when replacing defective parts.

3D printing is being used more often to create operational parts in reactors and infrastructure parts in nuclear power plants across the world. The examples above are but a fraction of current and new capabilities 3D printing is generating. Indeed, a team at Purdue University, Indiana, is also testing using AI as a quality assurance agent to aid in its printing of microreactors. This will lead to more cost-effective decision-making and data analysis in the development of future nuclear reactors. But other than structural and operational parts, what else can 3D printing offer the nuclear industry?

3D Printing and Nuclear Waste Recycling

Let's look at the controversy over nuclear waste, a long-standing criticism of using nuclear power. As I have stated before, one man's waste is another's resource. Back in 2008, I wrote an article called *The Waste of Nuclear Waste*. This was in reaction to environmentalist James Lovelock's statement in the Independent Newspaper (and his 2006 book *The Revenge of Gaia*) that he "would be happy to have the nuclear authorities build a concrete pit on my land and put some high-level nuclear waste in it. It gives off heat that could be used for hot water and central heating. It would be entirely safe and a waste not to use it." So, I thought if this was possible, and with countries considering underground facilities to bury nuclear waste, why should we bury the waste without considering alternate uses for it?

I wrote to James Lovelock via the EFN (Environmentalists for Nuclear Energy), a not-for-profit organization created in 1996. Initially based in France, they have spread worldwide with over 16,000 members and supporters in 65 countries. Later, I had a response from Bruno Comby, President of the EFN. He also commented that he would, too, "welcome a cubic meter or so of highly radioactive waste below my ecological home." However, he added that while the idea was good, besides the legality of it and the strict regulations regarding handling, transporting, and storage of the nuclear waste, the operation would not be economical, especially when compared to current gas and oil prices.

The main factor was that the amount of energy remaining in the vitrified nuclear waste (a process where nuclear waste is mixed with glass to immobilize it) would only be a very small fraction of the energy initially contained in the fuel, especially in reprocessed used fuel. While the low energy density in the waste, compared to the very high energy density in an operational reactor in operation makes it safer, the temperature is low (perhaps around 100°C). Thus, the energy from the nuclear waste may be only used for localized heat requirements, but not for widespread energy production.

As Lovelock also stated, nuclear waste has less potential of harming the human race or the planet, than CO_2 waste does. In not even attempting to put nuclear waste reuse to some good, then climate change will get worse and our solutions less imaginative and effective. We perceive nuclear products as wholly untenable until we have to use its power and life-saving medical applications. Currently, radiation is also commonly used in medicine and research, whether to diagnose and/or treat illnesses and diseases, or even to kill harmful bacteria in food. We use radiation more often than we think.

So, where does that leave 3D printing and nuclear waste? Well, just like in Chapter Six and my recycling ideas and Chapter Nine's energy harvesting concepts, can 3D printing find a niche here? And why would we want 3D printing involved with nuclear waste? The main answer is as you have just read above. 3D printing will allow the nuclear industry to save time, money, supplies and resources, manpower, and bring innovative new designs to facilitate the above, and all within safety parameters.

My initial thoughts for my 2008 article assumed that a suitable containment and heat-to-electrical transfer system (like a pyroelectric structure) could use nuclear waste to heat water or even power a house, neighborhood, or industrial building. There would be safeguards and monitoring systems to ensure that all health and safety aspects against containment leaks, explosions, fatal diseases, or fuel thefts were in place. However, as I have learned above, this may not be such a simple task. This also isn't a question of whether a 3D printer could run on nuclear waste energy; why would we want that? Rather I'm seeking to find a way 3D printing can genuinely help usher in a new age of nuclear energy from its waste.

In the UK, the Managing Radioactive Waste Safely (MRWS) guidelines advise on suitable methods to store reactor waste. However, their approach is the traditional, 'bury it safely and see if future generations and technology can deal with it'. In the US, proposed deep repository storage facilities, like the proposed Yucca Mountain Repository in Nevada, will be used for nuclear reactor fuel and other radioactive waste. But its protracted commissioning underlies the fears people still have over environmental concerns and the longevity of the radiation.

Nuclear waste would help mankind by doing what it does best: decay. Using waste fuel would also cut the need for 'nuclear miles', long-range transportation of nuclear waste to storage facilities, reprocessing depots or unauthorized dumps. Nuclear waste would also lessen our dependence on oil, coal and other sources of polluting energy. Now, the problem with this scenario is that the uranium half-life is measured in millions and billions of years. It's a slow lethargic process (compared with radioisotope thermoelectric generator (RTG) sources Plutonium, Strontium, Curium, and Americium), hence why uranium is not used for RTGs on space missions. Uranium's energy isn't strong enough for long enough.

So, this is my idea. Instead of having concentrated areas of stored nuclear waste, a series of small, dispersed sites could provide power to numerous industry and residential areas. The killer application could be as a power source at carbon sequestration plants, so that two waste product industries are more beneficial than one. Another option could be for portable contained nuclear waste units to be transported to regions requiring emergency power (like a radioisotope thermoelectric generator (RTG)) where the uranium's energy could be put to good use, especially in hard-to-reach areas where electricity has been disrupted, after natural disasters, and where other portable power sources are not yet available. Such units could become the ultimate decentralized energy system for the 21st century. The dispersed nuclear sites could be converted into nuclear battery depots, utilizing stored waste sitting there doing nothing.

That would sound crazy, except for the fact that a few months after I wrote this, an article was published featuring ex-SpaceX Engineers building cheap, portable nuclear reactors. Their initial idea was to use the technology for future Mars colonies, but then they realized it could also be used as portable nuclear power on Earth. Bear in mind the previous collaboration between the ORNL and Ultra Safe Nuclear Corporation (USNC), who also started working with Desktop Metal, who are aiming to create commercially competitive micro modular reactors using Binder Jet printers and you have potentially tied in 3D printing with portable nuclear power.

The ex-SpaceX group, called Radiant, led by founder and CEO Doug Bernauer hopes to develop the "world's first portable, zero-emissions power source," ushering in relatively lightweight, cost-effective microreactors. Like my vision, they also seek to deliver power to remote areas, enable quick installation of new units in populated areas, and have both commercial and military applications. Their prototype microreactor will produce more than 1MW, powering approximately 1,000 homes for up to eight years. For portability, it could easily be transported by land, air, and sea, ensuring affordable renewable energy to remote communities and help reduce fossil fuel usage.

Radiant's microreactor probably won't use recycled nuclear waste for power, preferring to utilize its "advanced particle fuel that does not melt down and is capable of withstanding higher temperatures than traditional nuclear fuels."

Helium coolant will replace the traditional water coolant, to reduce the corrosion and contamination risks. So far, Radiant is testing its portable microreactor technology at its Idaho National Laboratory (INL). Bernauer saw the gap in the market, observing that the current development of microreactors focused on fixed locations. With no established commercial system built yet, he hopes Radiant will be the first to do so.

Will such microreactors use 3D printing in their development and manufacture? Can they be powered by nuclear waste? Such reactors could provide clean, safe nuclear power and provide an alternative for remote environments. Can we repurpose used nuclear rods? Typically, these rods measure around 1 cm in diameter and approximately 3 to 4 meters long. They are bundled together into a fuel assembly. Upon their service use ending, can the radioactive rods be converted into pellets, using the metal recycling process 3D printing uses? What purpose would the recycled pellets serve if not a small battery-type function? I am sure there will be a legitimate shooting down of my ideas, but I hope there will be a few rebels who will focus on the creativity and have radical ideas of their own that will come to fruition.

Sooner or later, an entrepreneurial company or industry will realize that nuclear waste is a worthwhile commodity as a new energy source in unique locations (e.g remote, underwater, or in space). With global stocks of nuclear waste increasing, this new nuclear business could be classed as a renewable enterprise, with competitive prices and multiple uses, which could also be taxed much the same way carbon is today. The dreams of nuclear fuel being useful at the beginning of its life and could now be just as important at the end of its reactor life bringing cradle-to-cradle nuclear energy. If nuclear waste can be turned into a phoenix industry then that would enormously mitigate climate change impacts. Our future could be bright. It could be nuclear.

And this is where 3D printing comes in. I think 3D printing will be able to create and develop new prototypes, equipment, materials (whether based on vitrified glass, lead, or newer radiation-resistant fabrics), and complex geometries to be able to handle the radioactive heat and radiation levels. With residual uranium lasting millions of years, we should be able to wring every last drop of energy out of it for thousands of years even with ever decreasing energy levels. With the waste energy systems I have outlined in this book there has to be a way to capture the vestiges of radiation and heat, just like solar panels or even carbon capture techniques.

This is not a technology issue, but a cultural mindset issue. We've been told how to think about nuclear energy as being all bad. But with new applications of exponential technologies, like AI, robotics, adaptive 3D printing, and the metaverse (which could virtually simulate and study 3D printing in the nuclear realm), we may have to rethink our resourcing and recycling priorities. We've been told that spent nuclear waste has to be buried, but we have to work smarter not double-down harder on this problem. The waste will still be held on secure sites in secured containers, but we would be siphoning off the heat and absorbing radiation onto materials created through 3D printing.

We could use AI to oversee the process and use 4D and 5D printing to create containers and energy capture systems that react to the nuclear waste heat and radiation. There are currently sensors and radiation badges that alert one to radiation leaks, but with 3D printing materials you could add varying dimensions to the alert system:

- The secure containment could act as one big radiation sensor changing aspects to reveal radiation leaks.
- The printed material's shape and structure could alter to either release or contain the waste energy like a tap.
- The containment shell could turn color to indicate power levels or leaks (perhaps using a traffic light system with red for dangerous, amber for intermediate, and green for safe).
- The containment body could change size to regulate energy levels.

For power transmission, imagine wind turbine-type structures powered not by wind, but by radiation, sending electricity to homes via power plants. But the turbine rotors or radiation capture system may need to be 3D printed to create unconventional shapes to compensate for its radioactive load. We could also print structures to filter out radioactive elements from the air around the waste containers to further prevent widespread breaches and leaks. In the event of emergencies, tools and equipment could be printed from remote centers in time for emergency services to arrive.

With such ideas, this is why I believe the nuclear power industry aligned with 3D printing will have a cultural effect on and be beneficial to everyone. And if we're thinking of its effects in nuclear fission power, then there could be scope for 3D printing in the ever-closer fusion power industry, especially with its complicated torus-shape. This will be a 'watch this space' reminder for when you re-read this book in 10 years' time and note my ideas weren't so crazy. And I'm sure that if this particular prediction happens, then someone in the 3D printing or nuclear industry will win a prize for it.

CHAPTER ELEVEN

3D Printing X-Prize

A Vision of the Future

My vision is to have 3D printers in every household, workplace, and academic center, as ubiquitous as TVs, laptops, games consoles, microwaves, and soon, VR goggles, and metaverse interfaces.

I want to make 3D printing a lifestyle just like watching TV, listening to or streaming music or films, or gaming. To do that we have to bring the 3D printing profile up to par with these other technologies. How? First, by elevating 3D printing's publicity by advertising on TV or online and/or buying/renting from a highstreet shop. Everywhere you look you see every type of technology advertised in mainstream media, except 3D printers, which is still stuck in the realm of nascent niche tech. Also, it is still perceived as a stand-alone machine that cannot be used in conjunction with other tech.

Secondly, 3D printers need to have a user-friendly interface and be ready for the mass market as plug & play appliances. You don't need to know how a computer, TV, or microwave works, so why should a 3D printer have that disadvantage? One idea would be to preload 3D printers with catalogs containing thousands of items to print at low cost or to have an interface with a 3Dapp. Most people would just want to use the app to image, print, or buy objects with no in-depth knowledge of designing or slicing software or CAD training.

This is what needs to change. 3D printers have to evolve. They have to be equipped with hardware to physically connect to other equipment or software or wireless tech to enable accessibility and connectivity with other smart devices and adjacent equipment. 3D printers will be printing food so will have to be an economical device in the kitchen, be a tool/utensil maker for gardeners or DIYers, be child-friendly for school projects, be voice or touch-activated, and more.

There should be in-shop showrooms demonstrations and selling 3D printers with price ranges matching laptops and tablets. These printers can be sold as assembled or unassembled, with optional free assembly and set-up instructions by technicians once delivered. Added value will be through sales of spares/tools and the free 3Dapp catalog.

Detractors will tell you it's impossible, without thinking about how to solve those issues themselves. But the biggest issue will be the user interface. However, as described above and just like a self-checkout in shops where you can look up items on the product/grocery list screen is self-explanatory, each 3D printer will have an in-built catalog of useful and everyday objects that can be 3D printed. Indeed this would herald in the smart 3D printer. Going further, AI will be involved so all you would have to do is say 'Alexa, please print me a cup' and the AI interface will instruct the printer to make it for you as per your specifications.

XPRIZE

But what better way to raise awareness of the advances in 3D printing or to create new innovative uses for 3D printing than financial incentives? And there's no bigger prize than the XPRIZE. Founded by Peter Diamandis in 1994, his goal was to focus the public on "radical breakthroughs for the benefit of humanity" with "Revolution through incentivized high-profile competition." This was a multidisciplinary endeavor to foster innovative ideas and technologies toward solving the world's greatest challenges.

The first XPRIZE awarded in 2004 was the Ansari XPRIZE for space exploration research and development technology, with the USD 10 million awarded to

Mojave Aerospace Ventures for their SpaceShipOne, led by aerospace designer Burt Rutan and his company Scaled Composites. Licensed by Richard Branson, SpaceShipOne became the birth of Virgin Galactic.

As of January 2018, there are only seven completed contests and eight active contests, with many more being requested by the XPRIZE foundation. So would my series of 3D printing ideas discussed below be eligible?

The Bridge is an XPRIZE publication espousing radical ideas and inspiring others. I put together a letter below (which I have since updated) and contacted the XPRIZE through the Bridge at thebridge@xprize.org to upload my 'crazy idea' based on ideas mentioned in this book.

Hey XPRIZE,

I have 3 crazy ideas regarding 3D printing and think it's the future.

I am writing a book called The Future of the 3D Printing Culture. And since 2033 marks the nominal 50th anniversary of invention, I have come up with ideas to advance 3D printing within the decade. These ideas relate to energy and environmental issues with 3D printing and with producing a one-stop kind of 3D printer for its 50th anniversary.

I would like to add your response to my book, with your permission, whether I am successful or not. So I hope the ideas can kickstart XPRIZEs.

XPRIZE for:

1.   Scavprinting & Cryptoprinting
I created a couple of terms to connect 3D printing with energy harvesting, cryptocurrency, and cryptomining:

Scavprinting describes the general use of scavenging waste energy to power 3D printers.

Cryptoprinting describes harvesting waste energy from cryptomining rigs for 3Dprintingpurposestocreatethegrowinghardwareandinfrastructure accompanying cryptomining. Further detailed explanations could be provided.

These will investigate:

- Using the waste heat and vibrational energy (rig noise at 70 dB to 90 dB per rig and rising per additional rig) from cryptomining rigs and equipment to run small turbines or generators to power 3D printers. You would have two exponential technologies reinforcing each other.
- Scavprinting could also resolve issues around the carbon emission issues raised about cryptomining with the captured heat reused and reduction in noise pollution.
- Piezoelectricity can also be used for electricity generation using vibrations from industrial machinery, depending on the source of the force load, whether it's footfall in 3D print farms or cryptomining farms with workers walking up and down the floors carrying out monitoring and equipment maintenance.
- 3D printers can print heat sensors such as pyro/thermoelectric sensors to collect, distribute, and store scavenged heat through a pyroelectric energy harvesting generator.

The competing teams would have to practically demonstrate the ability to operate a 3D printer and print an object without a mains supply using only harvested waste heat/noise energy for an hour. The ability to do so would be via pyroelectric, thermoelectric, piezoelectric, or a newly invented means. The team with the best demonstration would win.

2. Skyprinting or Carbonprinting (CO-printing)

The harvesting of carbon sequestration techniques for 3D printing purposes.

Instead of using carbon capture to create highly-concentrated CO_2 for conversion into biofuels or carbon-neutral materials or to store as carbonate rock, why not convert a proportion of the carbon into carbon fibers or materials like polymer powder or filaments fit for 3D printing?

Excess carbon dioxide is a great waste product waiting to be taken advantage of. Let's not squander this opportunity to make a Global Warming villain become an industrial 3D printing hero resource for humanity.

The competing teams would have to practically demonstrate the ability to harvest CO_2 (preferable) or use a supply of sequestered CO_2 and convert into 3D printable materials. The material would then be used to print a carbon product. And the material will have to be quality-assured as comparable to current carbon printing materials. The team with the best demonstration would win.

3. The One-Stop Printer/Modulator/Creation Hub

The ultimate goal would be to see 3D printers in every household like a TV or microwave, sophisticated enough to be a plug & play appliance with apps and uploaded smart AI catalogs for household members and office staff to choose preloaded designs and products or to make their own designs.

The Creation hub would be ideal for printing, cooking, and fabrication of products whether in domestic, academic or commercial environments.

The main issue is the user Interface:

- How to make the 3D printer a plug & play appliance?
- How to create a smart, intuitive, and tactile 3D printing smart screen with menus, categories, drawing settings for AI interpretation and design?
- How to incorporate spoken commands, file sharing, etc?
- Incorporating features such as user-friendly buttons on the control panel: Green for Print/Go, Red for Stop/Cancel, Yellow for Pause/ Delay.
- How to enable multiple materials to be printed seamlessly?

For the latter interface criteria, perhaps there will be fitted:

- Integrated internal cassette-like modules which can be inserted in the 3D printer (instead of filament spools or pellet/powder pans) to input the different printable materials. (i.e. From printing a plastic cup and 5 minutes later printing a meatless steak on the same machine just by switching a print cartridge).
- Or rear-mounted refill canisters with pop-tops to supply feeders and nozzles without interruption. There would be canisters for plastic pellets, powder, or strips, metal powder, wood, food, ceramic, etc.
- The nozzle assembly for extrusion of the material should contain a multi-nozzle mounting, much like a microscope, with nozzles on a rotating mount and labeled for different materials, such as plastics, metals, and food, etc. The feed from the separate material canisters or cassettes would travel through a central nozzle assembly then through the appropriate nozzle as it is rotated into position to avoid cross-contamination.

Maybe we won't be using dimensional descriptors for printers in the future, whether 2D, 3D, 4D, or 5D, but calling them 'modulars', machines able to print/build anything from two to five dimensions in modulated units from any source material, almost akin to Star Trek's replicators/matter synthesizers and future programmable matter.

Within 10 years (2033—the 50th anniversary of the invention of 3D printing) the competing teams would have to practically demonstrate the ability of the all-in-one printer, successfully printing different materials within minutes of each other and demonstrating a practical and easy-to-use menu screen. The printer would be a little larger than a microwave and be able to fit on a kitchen countertop or desk and be considered a white goods appliance. The team with the best demonstration would win.

Thank you for considering my ideas for XPRIZE. I look forward to hearing from you soon.

I was then elated when in early March 2023, I was contacted by XPRIZE thanking me for my interest in the XPRIZE Call for Future-Positive Ideas. I was then offered the opportunity to complete their full submission form due in April with results back in June 2023.

Now, I had over a month to consider a more thought-out project, but I was too excited and completed the form overnight with ideas I could have fleshed out for maximum effect. So, before the deadline in April, I returned the forms and awaited the results like a kid waiting for a Christmas present. Whether I succeeded or not, it was still a great honor to have my ideas recognized as something that could potentially be developed for the benefit of both 3D printing and the world.

The Future 3D Printer

You may still be asking yourself why try to create an all-in-one 3D printer for a so-called creation hub in the home? Well, I have no doubt that in the future we will have 3D printers in the home as an appliance, as described in this book. Life is changing. It's becoming more flexible with transitions in work and technology blurring the lines between traditional and digital industries, between manufacturing and service industries, between commuting to work and working from home. 3D printing won't be just about printing and selling things, it will be about creating your own world, how you want it, when you want it. It will be a cultural game-changer worthy of an XPRIZE.

But more than that, pushing ahead to facilitate a one-stop-shop 3D printer will lead to more technological innovations such as smaller, smarter printers. Just like personal devices such as mobile phones and personal computers which were once considered futuristic devices, and then too big for practicality, until technology enabled miniaturization of hardware, massive internal file storage, portability, and lower competitive costs. So, we will have portable 3D printers to carry around whether at work, in remote locations, or for emergency situations. Our future mobile communication (comms) devices will one day be roll-up flexible or wearable tech which also function as combined identification, payment, and electronic key devices for various locations. Lose it, and instead of buying a new one you can print up a replacement anywhere, then download a copy of your information back onto it.

How will portable 3D printers work? First, such printers would use the usual materials including plastic, metal, silica, ceramic pastes, or perhaps specially-designed materials like composite gels, and biomass-materials etc, to create small objects on the go like the comms/ID/key/paycard device or clothing made from plant or biomass material or scrap plastic. And by then we won't be worried with hotends as portable and micro-3D printers will use room-temperature bonding methods. Such methods, dubbed direct or fusion bonding, are currently in use.

The technique utilizes the chemical bonds between two surfaces of any material depositing successful layers of joined wafer-thin surfaces through wafer bonding techniques. Silicon is the most widely used material for this process, known as silicon direct bonding, or silicon fusion bonding and is used in the manufacture of silicon-on-insulator wafers, semiconductors, sensors, and actuators. This may be pushing the definition of a 3D printer, but then again, 3D printers will evolve and new techniques will merge or diverge from others. Sounds impossible? Consider these two recent breakthroughs in 3D printing.

Metallic Gel for 3D Printing at Room Temperature

Scientists from North Carolina State University, Northwestern Polytechnical University, and Tianjin University have successfully 3D printed solid metal objects at room temperature. The metallic gel they created is made from a "mixture of micron-scale copper particles suspended in water and a small amount of a liquid indium-gallium alloy", paving the way to manufacture electronic components, devices, and products. The copper in the suspension is distributed evenly allowing the material to conduct electricity. But crucially, it also remains stable in the aqueous solution which maintains its form after printing. Once printed, the item can undergo the curing process, even at room temperature, experiencing chemical or physical property changes leading to greater solidification. Further, curing can also be affected by other environmental factors, such as subjection to heat, UV light, or chemicals, depending on the material used. And as I discussed in Chapter Nine, such transformative features are part of 4D printing, allowing for material movement over time. So, now you can imagine portable 3D or 4D-capable printers producing objects at room temperature. Currently, this has been accomplished using their metal gel. However, innovations will see new materials and applications developed, advancing 3D printing, and creating the conditions for 3D printing to become not just a household appliance, but even a personal device.

3D Printed Elastic Conductor Materials

A research team at the Korea Institute of Science and Technology have created 3D-printable elastic components that can conduct electricity. Their process relied on producing a special emulsion-based composite ink, consisting of liquid components diffused within a conductive elastomer, a rubbery material which can conduct electricity. With its resulting structural integrity and flexibility, precise complex geometric items can be 3D printed allowing for the creation of stretchable solid-state elastic conductors with electronics. In fact, the research team demonstrated the technology by using the ink to successfully "create a wearable temperature sensor with a stretchable display". The team is confident their ink will allow for "omnidirectional printing of elastic conductors," advancing the field of wearable technology. Such materials would certainly have a cultural impact on our everyday lives.

Metallic gels and composite inks, along with other developing 'exotic' materials could form the basis of a new material reality for 3D printers. With material technology advancing every day we will have products that can be broken down into gels and inks and reconstituted as new products by the 3D printer. So instead of carrying around solid materials for printing we could carry pouches or flasks of composite gels and inks for instant printing on the go. Today, we don't think twice about carrying a water bottle, hand gel dispenser, or a tablet or laptop around. The portable printer would be as simple as traveling with those components of our current material culture. The printed product could then cure in the sun, or react to the air (especially with CO_2 or other greenhouse gases) or drops of water to fulfill its function.

Back to the Future 3D printer

With the above potential technology in mind, what would a portable printer look like? There are already small compact palm-sized 3D printer models currently available and powered by battery or solar power. However, the new portable printers will be even smaller and adaptable made from 4D or 5D printed materials enabling a flexible structure, intricate designs and prints for a just-in-time printing capability. My vision is that it will be tablet-sized, able to fit into a pocket as a foldable device. The bottom foldable surface would be the print bed and the top surface fold up perpendicular as a menu screen with control functions. The menu screen would have a design feature linked to your mobile app for pre-loaded or saved necessities like comms, clothing, utensils, wearable tech, and jewelry, etc. An extension arm on the top surface would pull/fold out acting as the material input/producer, print head, and extruder. For 4D or 5D printing, there would be multiple fold out arms, resembling a rotating grapple, with the product printed from its arms in the middle of it. The printer would run on batteries and/or solar power or even be linked to your wearable piezoelectric recharger, powering up as you move.

Why would your flexible comms/ID/payment/key device be 3D printed? Because of everything that I have discussed above. Your device (and many others) would be customizable for you possessing attributes such as:

- Variable printing techniques (e.g. 4D or 5D printing) internally or externally adding unique security features.
- Variable material components (e.g. recycled, gels, inks, metals etc) for flexibility, recyclability, durability, and allowing seamless and multiple electronic functions.
- Ability to be printed in non-centralized factories, at home, on-the-go, at anytime.
- Be a wearable technology fashion accessory.
- Advance the cultural validity of 3D printing.

This is the world I am thinking of. It's not magic, but a technological process, driven by a cultural need. If that sounds good, then that's the next challenge for you to resolve.

In the meantime, on June 1st, I duly received my results from the XPRIZE Ideas Team:

Dear Raymond,

Thank you for your participation in the XPRIZE Global Call for Future-Positive Ideas. We regret to inform you that your submission has not been selected to advance.

The field of submissions was highly competitive, and our team of experts evaluated each submission with careful consideration. Although you presented a promising idea, our expert evaluators have determined that your submission unfortunately did not meet the criteria for an XPRIZE competition at this time.

We are sincerely grateful for your participation, and we encourage you to remain connected with XPRIZE through our blog or by following us on facebook, instagram and twitter to be informed of future opportunities for engagement. We wish you all the very best in your future endeavors.

The XPRIZE Global Call for Future-Positive Ideas Operations Team

Of course, I was gutted, but I was immensely proud and honored to have been considered for the competition in the first place. The words I keyed in on were that my ideas did not meet the XPRIZE criteria at this time. So, it was left open that my idea could be further developed into a more promising concept either by myself or others to satisfy the conditions and re-enter the competition. So, if you are up to the challenge, why not revisit the problems I tried to solve, with fresh eyes and ideas, and create an XPRIZE for the future of 3D printing.

However, in early May 2024, I completed a short XPRIZE Design course hosted by Peter Diamandis. In it, I learned why some prize topics were not as good as others and what attributes were key to designing a great XPRIZE. When I applied them to my 3D printing XPRIZE idea, I realized I had missed the mark and underestimated the potential advancements of 3D printing.

The main attributes among the 18 Diamandis listed, I had not accounted for were:

1. Was the field stuck? Was the invention going to happen anyway?

2. What was the finish line? Would it be obvious? Would people care about the story?

3. Is there a telegenic finish? Would there be public attention?

In answer to those, while I believe the field is somewhat stuck in a print-sell model, the invention of the one-stop-shop 3D printer will probably happen before my nominal 10-year prize deadline. Unlike sending a spaceplane to space, would people care about the race to create an all-in-one printer? What is the finish line – printing out a food dish in between a metal and plastic printed item with no contamination or printing multiple elements over days or weeks? And lastly, 3D printing can be boring to watch. A recorded TV or online show can create a montage, but live, during a competition, would the world watch a metal drone rotor being printed followed by a chocolate cake and then a plastic costume helmet?

So, now, as a domestic appliance, the all-in-one printer becomes a problem for the commercial industry to solve and release as a product, rather than utilizing time and resources on an XPRIZE. This isn't to say a 3D printing XPRIZE couldn't be created, but it would have to meet the key attributes for a great prize, which unfortunately, my idea did not.

3D Pioneers Challenge

And of course, the XPRIZE is not the only technology competition in town. But we need competitions related specifically to 3D printing. The International Competition for Additive Manufacturing and Advanced Technologies established by the 3D Pioneers Challenge is in its 9th year. Its goal is to pursue long-term sustainable and creative innovations and operational projects that will advance the additive manufacturing industry. With collaborative initiatives in mind, the competition will focus on medtech, design, fashion, mobility, and architecture technologies in the digital, electronics, machinery, and material realms in a quest for sustainability in the 3D printing realm.

While there is no literary category, I sent the initiators of the 3D Pioneers Challenge, Simone and Christoph Völcker, a request to see if I could submit this book in the competition. I figured that with all the innovative ideas, I could introduce new concepts and inspire others. While I did not submit at this time, perhaps in the future there will be prize categories for 3D printing literature.

There will also be other opportunities. Sites like 3DwithUs, Cults3d.com, 3ders. org, Pinshape Contests, MyMiniFactory Challenges, GrabCAD Challenges, Instructables Contests, CGTrader Awards, Makeable, and many more offer prizes for 3D printing designs and prints. So, get your thinking and printing caps on and use the cultural customs of competition to help push the industry forward. Good luck.

CHAPTER TWELVE

The Cultural Future of 3D Printing

What's in a Word?

Language is an integral component of culture. Many words and terms for technology have been simplified over the years and slipped into informal vernacular and slang. To effect an easy transition into society, 3D printing should follow this model. Additive manufacturing is such a case. Yes, it accurately describes 3D printing, but it is much too prosaic a term for general usage and is a commercial and industry-based term. Even 3D printing does not easily roll off the tongue and can still be a little more user-friendly.

The term '3D printer' is also problematic as it describes both the machine and the person. I'm sure you've noticed that in the text but understand the context it's written in. I can say I'm a 3D printer using a 3D printer to 3D print. Some people use the term 3D printerer, but that looks and sounds strange and my spellcheck corrected it a few times. And I'm sure we won't go down the 3D printerman/woman/person route!

So, what should we call the act of 3D printing in the future? We need a single simple word, not words and terms, that turn 3D printing from a gatekeeper technology where only insiders are privy to the knowledge into an accessible user-friendly technology for all. Let's look at some possibilities for the future 3D printing lexicon:

• Thrint

Fabbing and Maker are also terms related to 3D printing but not exclusively. I thought I had found a unique word with *thrint*, but when I looked it up it was already in use with two sources noted in Definitions.net and the Urban Dictionary. Both sources state:

> Thrint – verb meaning to 3d print an object on a 3d printer. [For example] I was easily able to thrint a new part for my classic car. Etymology: a contraction of 3d print.

This was submitted by brettlarenatkins on June 26, 2021, in Definitions.net and by Souty also on the same date in the Urban Dictionary with a similar example of "I was easily able to find a file to thrint a new part for my car." Same person? If so, they are the originator(s) of the word in the 3D printing context, at least in the public domain.

There is also an existing 'vacant' thrint.com website with the express purpose of being for 3D printing, just waiting to be bought and used by someone, as of this writing. So use thrint, publicize it, and associate it exclusively with the act and art of 3D printing.

December 31st, 2022, I went further and added the comment that thrinting, thrinter, and thrintable are associated words to the Definitions.net. Maybe someone taking an apprenticeship course in 3D printing is on a 'thrintership'. I decided to keep on using Definitions.net and the Urban Dictionary to build my 3D printing lexicon.

• Thrab

If you are a fabber then thrab, thrabber, thrabbed, etc.; are also possible, as I defined thrab on the Urban Dictionary and Definitions.net as:

Verb: Of and relating to 3d printing and fabber (digital fabricator) manufacturing.

Sentence usage: I will be able to design and thrab the product for you.

Etymology: A contraction of the terms 3d printing and fabber. Associated words thrabber, thrabbing, thrabbed, thrabable.

Thrint and thrab may look and sound strange but most new words do. Plus when someone says they're going to print something up it still has connotations of 2D printing, so we need to differentiate between 2D and 3D printing and incorporate a cultural value to the word.

- **Scavprinting and Cryptoprinting**

Maybe I should add Scavprinting and Cryptoprinting to the online dictionaries so the development of the 3D printing cultural growth can continue.

Define Scavprinting for the dictionary:

Verb: the general usage of scavenged waste energy (thermal variation and acoustic vibration) from any source or equipment to power 3d printer(s).

Sentence usage: The factory was the source of the waste energy for scavprinting the parts.

Etymology: Contraction of the terms scavenging and 3d printing.

***

Define cryptoprinting for the dictionary:

Verb: To use harvested waste energy (thermal variation and acoustic vibration) from cryptomining equipment to power 3D printer(s) for product manufacturing.

Sentence usage: The products are being manufactured through cryptoprinting.

Etymology: Contraction of the terms cryptomining and 3d printing.

Maybe those words will catch on. Or if you call yourself a maker with a 3D printer, that would bequeath thrake, thraker, etc.; or if you want the additive manufacturing angle then try thrad, thradder, and thradding.

Whatever term you use, it would give 3D printing a specific cultural word, easily recognized anywhere.

Would we need such words for 4D printing—say, Frint (for Four-D printed) etc., or for 5D printing—perhaps quint? That would line up with thrint and make it a grouping of terms for 3D printing:

- Thrint – 3D printed

- Frint – 4D printed

- Quint – 5D printed

Or it could be made simpler just by adding a capital letter 'D' to the end of three, four, and five and say it as one word, like the past tense of the word (liked 'freed' or 'floored' or 'fined') so:

- ThreeD

- FourD

- FiveD

- **3Der**

The above may fit in with the term '3Ders' (or 3ders) I've seen individuals and companies use online. So, during July 2023, I 'hard' rubber-stamped them into Definitions.net and the Urban Dictionary to add to our 3D printing lexicon for users.

Define 3Der (also 3der) for the dictionary:

Noun: An individual or organization that practices 3D printing or is active in the 3D printing industry.

Sentence usage: The 3Der will design and print your prototype products for you.

The 3Der industry is surpassing the traditional manufacturing processes.

Etymology: A contraction of 3D printer.

So you could say you're 3D'ing a model or you've also 3D'd a product. So of course, we'll be 4D'ing and 5D'ing objects in the future. That sounds more sensible. Such a grouping above would make it easier to say and describe.

- **Throdel**

In fact, we could also contract the words 3D model into 'Throdel'. I have defined this term in the Urban Dictionary and Definitions.net as:

Noun: A 3D model or 3D printed model.

Sentence usage: This is a scale throdel of the building we designed.

Etymology: A contraction of the words 3D printed model.

• Thratter

As I have mentioned Star Trek's matter synthesizers a few times we can also come up with a term for 3D printed matter: 'Thratter'. I have defined this term in the Urban Dictionary and Definitions.net as:

> Noun: 3D printed matter, materials, or products.
>
> Sentence usage: All the aircraft parts are composed of thratter.
>
> Etymology: Contraction of the words 3D printed matter.

But Thratter can also be a verb, as defined as:

> Verb: The act or process of creating or producing 3D printed matter, materials, and products.
>
> Sentence usage: We will thratter the car parts for you to ease prototyping, costs, and sustainability.
>
> Etymology: Contraction of the words 3D printing matter.
>
> Associated words: Thrattering, Thrattered.

Perhaps thratter could be further contracted in the vernacular to Thrat or Thratt and still retain its meaning. And maybe instead of a generic synthesizer we could 'Thrintesize' or 'Thrattersize' objects to clearly denote the synthesizing of 3D printed items. Whichever term we use it would add to the dictionary of 3D printing culture.

Of course, I'm a middle-aged man and maybe my proposed words are too derivative, traditional, and outdated already. So, maybe the youthful readers can create new words or phrases for their generation; TikTok-up the 3D printing industry and culture and give 3D printing the social media presence it needs.

But, also, with the rise of alt realities and AI, there is one more 3D printing term to invent and define:

• Metaprinting

As you read in Chapter Nine, I created the term metaprinting. Advantages? Even faster and more accurate than a digital slicer with the ability to 'physically' manipulate the 3D printer or object in the alt-reality. AI intervention will optimize decisions on 3D printing setups, design variations, and print platforms. Multiple printer types, print set ups, and prototyping prints can also be created simultaneously within seconds. With your chosen configurations you can then print the results in the real world and/or share and print your files over any medium to any connected printer in the world. You could keep the files in the Metaverse/alt realities creating a metaprinting community.

I defined the term in the online dictionaries as:

> Noun: The use of 3D printing technology in the real world in conjunction with the Metaverse, VR, AR, other alternative reality mediums, and/or AI for the testing, developing, and prototyping of designs, actions, and other usage in those alt realities with translation back into the real world for 3D printing applications.

> Sentence usage: We will be metaprinting multiple product variations before real-world printing the best results.

> Etymology: Combination of the root of metaverse (as the umbrella term for alternative realities) and 3D printing.

> Associated word: metaprint.

The 3D Printing culture could spread in the metaverse creating a new shared medium to grow the industry. And looking ahead to moon and Mars missions, metaprinting may become the most efficient use of resources especially if coupled with AI to design and create new rockets, space-worthy tools and equipment, food and power infrastructure, satellites, spacesuits, and even bases.

In fact, The Mars Society, dedicated to manned missions to Mars, announced their forthcoming initiative to create the Mars Technology Institute (MTI). Part of its purpose would be in developing the technologies needed for infrastructure to settle Mars. The Mars Society's director, Dr Robert Zubrin has called for investors and inventors to achieve this goal advocating the necessity of such an organization now with a long-term vision to colonize the Red Planet. While the MTI will initially focus on biotechnology (e.g. genetic engineering, microbial food production, aquaponics, and synthetic biology), Zubrin hopes that automation, robotics, and AI will also follow suit once their technology matures and provides patenting and licensing opportunities for the inventors.

And as described in this book, 3D printing will be a great asset in these technological endeavors setting up a Martian cultural tradition. The Mars Society runs several analog research stations where they can test 3D printing, especially metaprinting methods, testing potential equipment, materials, and infrastructure for colonies. Valuable knowledge for would-be Martian astronauts, researchers, and entrepreneurs could be gained very quickly. So, to the future MTI and Martian colonists, get printing!

Beyond 3D Printers

Maybe we won't be using dimensional descriptors for printers in the future, whether 2D, 3D, 4D, 5D, or 6D but calling them 'modulars', machines able to print/build anything from two to six dimensions in modulated units. Instead of paper, we'll have smart flexible printed screens created by these modulars able to build anything from any source material with synthesizers or by programmable matter, just as in Star Trek. But while we are expanding the 3D printing industry's culture, what will happen to many manufacturing and service industry jobs affected by the 3D printing revolution?

3D Printing Manufacturing Outlook on Culture.

Awareness of the 3D printing industry is growing around the world as are sales of printers and products by individuals. However, while I have been espousing the materialism and lifestyle of the 3D printing culture, we also have to spread 3D printing into non-industrialized, non-centralized, and non-urban cultures as opposed to furthering globalized countries' monopolization of yet another technology industry. The US, China, Europe (mainly Germany, France, and UK), and South Korea dominate the manufacturing and sale of 3D printers and subsequent product sales. We need outreach programs to implement online and practical courses, with alt-reality training in developing countries and tech-poor areas to lead to a real sustainable, long-term global 3D printing industry. Nobody should be left behind in the 3D printing revolution.

Countries in Africa, South America, Oceania, South-east Asia/Asia Pacific Region, and other areas not already fluent with 3D printing or without great manufacturing and/or export bases could use their own domestic resources to become more self-sufficient through 3D printing services. How? Here are a few ideas:

- They could become hubs for remote 3D printing farm services, just as other countries have established cryptomining farms, and IT and call center services.

- Like China and Dubai, they can establish dedicated learning centers and certifications for 3D printing, encouraging exchange programs for academics and entrepreneurs.

- They could start producing the 3D printing adjacent products and merchandise, I have described throughout, such as textile factories creating fashion for 3D printer aficionados.

- They could establish 3D printing art galleries and exhibitions, collecting original 3D printed artworks or reproductions of famous works.

- If we return to the recycling discussion, for instance, it is a well-known practice for developed countries to contractually (and sometimes illegally) dump rubbish in lesser-developed countries, who then process that material to their own ends. With the plastics and metals involved, these countries could learn how to process the waste into 3D printing-ready material, whether filaments, pellets, or powders, and sell it back to the rubbish donors or develop their own production service, dependent upon quality assurance reviews. The infrastructure could be built up over time, but the waste we pile up in other countries could create a niche and prosperous 3D printing recycling industry for themselves.

These different countries, communities, groups, and individuals can use 3D printing at their own pace and for their own purposes, expressing their particular culture through 3D printing. We can then diversify the culture avoiding a monoculture in 3D printing which would surely stagnate the industry. Manufacturing, sales, distribution, and usage of 3D printers must be a global mission allowing the 3D printing industry and culture to thrive. 3D printing must also become a bottom-up industry not a top-down one dictated by corporate or commercial interests. This goes hand in hand with education, government promotion of 3D printing, grassroots learning, and the adoption of 3D printing by more diverse industries. With these, the successful future of the 3D printing culture can be assured.

And if you really want to get crazy with 3D printing ideas for other cultures, let's look at possible 3D printing opportunities in relation to marine and desert cultural lifestyles. Why not have ship/sea-based 3D printing facilities?

Organizations already collect polluting plastic from the seas and oceans, but with all the amount of rubbish dumped in international waters, especially microplastics, we should have the capability to collect and process that material at sea. The plastic (micro and macro) waste could be recycled and transformed by so-called sea-printers into useful 3D printing material on site to create more printing materials, artificial coral, sea habitats, coastal defenses, fishing nets and equipment, cabling material, ship equipment and parts, rescue equipment, oil spillage collection containers and devices, and more. Or, with some fish stocks and fishermen livelihoods under threat, an alternative income for them could be in collecting sea-borne plastics and selling them to 3D printing recycling companies.

Sea-printers may also have a role in desalination plants, processing the dried brine left over from its separation from the water (with other chemicals). The recycled salt could be formed as pellets or powders and mixed with other materials to print solid structures (e.g. like existing salted concrete), create material blends and inks, and other products to sell since desalination can be an expensive process. Alternative salt product sales could help offset some of the desalination process costs.

And from the sea to the shores. With countries dealing with increasing desertification or wishing to utilize the abundant desert resource, the sand is literally a 3D printer away (well, with further processing of sand with 95% silica proportions) from being turned into products. Manufactured items could include: glassware, glass artworks, solar panel and solar tower parts and infrastructure, dew/mist/rainwater-collection tanks, greenhouse structures, water well and canal liners, glass bricks ('glicks'?) for lightweight structures and aesthetic effects, or maybe even high-tech screens and walls. (My own preference would be glass products for a combined vertical farm/solar tower farm). Instead of mining, processing, and importing the traditional earthen, concrete, and steel materials, under-utilized glass resources could be used as a viable substitute in some cases with 3D printing offering another revenue avenue for the construction industry and an environmental solution against sand encroachment.

In fact, one global company is seeking to invest in more opportunities using sand 3D printing: Tesla. The electric car manufacturer is looking to cut car production costs by creating molds through sand binder jetting. Currently, their innovative gigacasting process uses large presses to mold the front and rear structures of their cars. Sand binder jetting will now enable Tesla to sand 3D print molds and to die-cast almost the entire underbody of their cars rather than having to make over four hundred separate molds for the same underbody. This would significantly save on manufacturing costs and time in creating and using metal molds. So, the next time you're laying on a sandy beach or traveling in a desert, think of all the future electric cars that will become less costly with the use of printed sand beneath you.

It could be surprising how many opportunities and resources people, organizations, and countries actually and potentially have when it comes to 3D printing. It's a matter of perspective, adaptation, and action, the stuff of cultural growth throughout history. Considerate studies into such technologies and the right investments in funding, time, and social enterprise would be required. However, the cost of 3D printing's success across many traditional industries could come at a high cost to human labor.

3D Printing and Job Displacement

While I laud 3D printing and its opportunities, there are also potential downsides to its rising manufacturing preeminence, especially when more homes, schools, and businesses start owning their 3D printers. There will be less reliance on external traditional manufacturing services. This could lead to one of the biggest shake-ups in the economic sphere: mass job losses. We want the 3D printing cultural legacy to be positive not putting people out of work. Hence giving back in ways people can survive with 3D printing skills, experience, and the opportunity to benefit from it.

This vast cultural upheaval has been dubbed Technological unemployment (TU), with job losses expected through robotics, AI, automation, 3D printing, and other technological advances and processes. 3D printing could make some factory and manufacturing workers obsolete, resulting in disruptive market impacts. However, those exponential technologies will somehow have to give back. There won't be a free lunch for them. Accordingly, how can the 3D printing industry reward owners and users but also compensate displaced factory and manufacturing workers? One way would have to be to create new economic industries and methods to guarantee society is not critically disrupted by the new Industrial Revolution.

Coherent plans to compensate those who have been displaced by 3D printers and other non-human systems will be required. Current debates suggest governments, tech companies (including 3D print companies), economists, and employers will have to identify methods to protect or compensate technologically unemployed humans. How? According to the World Bank in their World Development Report (2019) they believe that "While automation displaces workers, technological innovation creates more new industries and jobs on balance."

In other words, for 3D printing, job losses could be mitigated if workers are re-skilled for 3Dprintingworkorassociatedwork, like design, engineering, materials manufacturers/sellers, product manufacturers/sellers, material recycling, etc. 3D printing could help free people from certain jobs or they can actually use 3D printers themselves and establish their own home businesses. 3D printing will become the new renaissance of creativity and industry for both individuals and communities. Thus, they could future-proof themselves against further TU job losses. 3D printing will create its own economic ecosystem which will require workers, even in automated services. While everyone displaced cannot be reemployed in the 3D printing industry, there are economic and legal experts who claim that companies should have a moral obligation to assist those who are affected by TU. While morals may not equate to legal understanding there should be adequate provisions considered.

However, what is required are government programs, investment, and the funding to provide training in this era of exponential technology, starting with STEM (science, technology, engineering, and mathematics) training in schools for the next generation of workers. On the downside, some academics argue that increased STEM training could lead to too many highly skilled workers, decreasing their value. But, factoring in the information that 3D printing start-ups and companies are always seeking designers and engineers in various roles and that 3D printing is not even out of its nascent phase then the demand for STEM training could be validated So, an enterprising worker affected by TU could invent a new kind of 3D printer or an application to expand the 3D printing universe.

Economic Solutions to Unemployment Through 3D Printing

Universal Basic Income

As 3D printing becomes part of the de facto new Industrial Revolution, there will be mass job losses, accompanied by further losses through automation and robotics. Society will have to change its outlook on how the service economy and traditional manufacturing work. Humanity will have to cope with fewer humans at work with less work to go around and how to keep people paid, housed, and fed with less or no income.

Though welfare payments or some form of government subsidies exist in some countries can they be repurposed for those displaced through Technological Unemployment? New social security, welfare and benefits, or pension schemes could be established to deal with TU, protecting workers from any other collateral technological redundancy. If not, an alternative basic income system called Universal Basic Income (UBI) may have to be established as a future payment system. UBI or similar systems would ensure people receive unconditional regular tax-free sums of money.

UBI trials have taken place in Brazil, Canada, Finland, India, Iran, Kenya, Namibia, Norway, South Africa, South Korea, the United States, and Wales with other countries also considering the payment system. So, if a 3D print farm directly displaced workers from a car or bottling factory then released workers would be entitled to UBI. Not all the UBI trials have been successful as UBI funding and payments can be complicated. However, that will not negate the need for UBI schemes in the event of mass TU with checks and balances to ensure fair and transparent operations.

Taxing 3D Printing

In a similar vein to proposed taxes on robots, as mooted in 2018 by Microsoft's Bill Gates, governments could tax a company's use of 3D printers, where it has displaced workers. Taxes could be raised by governments or the private technology sector should raise the funds to compensate their displaced workers. 3D printers could be taxed and insured, like vehicles, especially if it is an income earner for a corporation. And a proportion of the income from 3D printer sales, 3D printed product sales, and other revenue could go toward a UBI fund. Or a combined private-public partnership with governmental and tech company funding could pave the way for a unified UBI system.

Would such an implementation disincentivize companies from replacing too many workers with 3D printers, while at the same time providing a guaranteed living wage for its technologically unemployed workers? Governments and organizations would need to be clear on who would regulate the 3D printing tax system; the government or private companies? Either could raise and distribute the funds to varying communities and organizations or provide training for those TU-displaced workers.

So how will taxing 3D printing work? In theory, governments would tax a 3D printing factory or facility, raising funds for displaced workers and providing re-training. For example, if Ikea were to switch to 3D printing all their furniture displacing human workers, they would then be taxed for the printers and providing funds for UBI-type schemes. Taxing 3D printers would also help the government recoup the loss of income tax from workers who have lost their jobs to 3D printing.

Of course, details of such taxes need to be confirmed and sourced, whether from 3D printing-generated profits, 3D printing labor-saving efficiency, product sales, 3D printer sales, or a direct tax on 3D printers per individual unit. Just like the increased taxes imposed upon tobacco products to deter current and future smokers and to pay for health programs to assist ill smokers, these 3D printing taxes would act as a safeguard to TU employees' payments and protect against complete 3D printer replacements.

Other Economic Solutions

How else can you get paid by a 3D printer? With companies being taxed for 3D printing usage, individual 3D printer owners may face a lower rate of taxation (there being threshold limits on taxation on a sliding scale of ownership and usage). An individual 3D printer owner may earn a living through their 3D printer either as a business or by renting out their 3D printer(s), sharing with others, or using a communal 3D printer pool or a farm. Upon completion of a printing assignment, the 3D printer owner would receive payment tokens, crypto, or other currency from the customer. So, a passive income stream could be generated through continuous loans of 3D printers for tasks. The 3D printer would pay for itself. And an annual tax similar to owning a car or fleet of cars with annual insurance and registration taxes would be levied.

Perhaps a solution to limiting job losses by 3D printing would be to create new industries either solely for 3D printing or where 3D printing is practically needed instead of just enhancing or replacing traditional manufacturing jobs. Take the potential for jetpack racing for example, or biomedical advances, AI-enhanced 3D printing services, novelty food dispensers, or developing space and ocean exploration technologies. 3D printing can cut (and print) its own cultural path without having to totally disrupt the economic landscape.

Future-Proofing the 3D Printing Industry.

2033 marks the nominal 50th anniversary of 3D printing. Before then, it has to reach beyond the simple merchandise-selling model of previous tech industries. We are the service and ideas industry needed for the future. There are a multitude of adjacent possibilities available for 3D printing rising to an exponential rate in the future. 3D printing has to be more than relevant. It has to be culturally essential.

In some ways, I feel the proposed culture of 3D printing may be resisted by some of the 3D printing community who may feel the ideas of this book are impossible or who think 3D printing is advanced enough or that only the establishment can advance 3D printing technology. But this industry is open for all and the only limitation is in your mind. We don't want to keep 3D printing as a siloed industry. This insulative thought process only protects a small portion of the industry, the gatekeepers, and traditional manufacturers who want to maintain the print-and-sell market model to maximize profits akin to the oil industry profiteering before more renewables make them obsolete. The culture of 3D printing still has a way to go, and its development will be organic, but inexorable in motion.

Twenty-first-century lifestyles are becoming less traditional with people eschewing 9-5 jobs, careers for life, nuclear families, and hard currency economies, etc. 3D printing would be a method of plugging gaps within that changing cultural landscape offering the same print-on-demand options as in the literary printing world and the fluidity of digital file-sharing akin to the music industry. If the autonomous robot industry does develop to the extent of providing 24/7/365 delivery services, then 3D printing will provide the just-in-time/around the clock manufacturing base. If you can't print it yourself or can't wait for the traditional working hours delivery slot? Have the item 3D printed and delivered to you at 2am. 'Bespokeness', 'iterationality', and 'flexi-materials' may be the watchwords as 3D printing ushers in a seismic cultural shift in manufacturing and services.

I believe the scenarios outlined in this book, while sounding crazy, are not impossible to achieve. Well, as the sayings go, "If you don't ask, you don't get" or "If you don't try you won't succeed." We just haven't tried, thought of, or implemented some of these ideas to any small degree. We're not talking about room-temperature fusion or perpetual motion machines. We know we can build things molecule by molecule and 3D print in more ambitious styles. The fusion and synergy of technologies enabling a 3D printing revolution will just be the beginning. The stumbling block won't be the technology, but our imagination, creativity, and the cultural interface to ultimately fulfill the potential of 3D printing.

It's part of our cultural DNA to understand and manipulate technology. 3D printing can be the supreme expression of this, the progenitor of a new material culture leading to a new human civilization.

Let's create that future now.

CHAPTER THIRTEEN

The Story of 3djacent Solutions

As I mentioned in the introduction, I have included my business story here. This isn't an indulgent plug for my company, but an overview of my thought processes and actions in the inception and incorporation of 3djacent Solutions Ltd on 20 January 2020 and the intervening timeline until now.

Why am I telling my story as a non-successful businessman? Well, it gives me perspective, looking back at my successes and failures. It reveals how hard it is to keep going. It's the story of the invisible parts of the business, the emotional and physical toll taken trying to set a path even in a normal business, let alone trying something different. The business path is never straightforward but takes you backwards and sideways. I hope that with my sharing, you will want to take that step forward either by yourself or with others. Lastly, it also shows that while one opportunity evades you, other ideas and experiences are created and cherished. Hence this book on the conception of the 3D printing culture.

So, first off, to be transparent, my company didn't live up to my expectations. It's not a total failure as some salvaging could be achieved, but for now, it's not trading.

There were a few reasons for this. One main reason was that I wasn't my own great company salesman. I also feel the idea for my company was ahead of its time. I really do. Because, chiefly as any business person is warned not to do, I created a company that had no customer base or an existing problem to solve. The 3D printing culture and lifestyle I am projecting does not exist, yet. Which was fine. I was still going to create the first 3D printing lifestyle and management company. I was being very ambitious and forward-thinking, but I just didn't know how to do it by myself.

3D printing is not so advanced in mainstream minds and the media that a company marketing a 3D printing lifestyle could make an impact. So, how did 3djacent Solutions rise and fall?

The Diary of 3djacent

2019

I had related that I had taken a 3D printing course in May 2019, bought my first printer, and wanted to buy associated equipment to enhance my printer. But there was nothing beyond spare parts, tools, and filaments. I wanted the adjacent infrastructure like a custom-built table or workbench. But there was no such thing. So I was determined to make one, and that became the basis of the company—to make and sell things adjacent to 3D printing, hence 3djacent. And since I was not a printing and selling company, but offering solutions, I incorporated the company as 3djacent Solutions Ltd, with taglines such as:

- Your 3D Printing Solutions Company
- Managing Your 3D Printing Solutions
- Your 3D Printing Lifestyle and Management Company

I quickly created a landing page on the GoDaddy platform, in December 2019 to secure the company and domain name. And then did nothing with it besides adding the below, with a promise of 'Services Coming Soon':

Why 3djacent?
You want a 3D printing services and design company with products and solutions adjacent to 3D printing.

Why 3djacent?
You want to maximise your 3D printing potential, experience, and lifestyle.

Why 3djacent?
You want a company dedicated to pushing the boundaries of the ever-evolving 3D printing landscape.

Not much happened between the course in May 2019 and the creation of the website in December. However, at an entrepreneurs' Meetup, I was introduced to another 3D printer who was mooted as a possible business partner, since I was looking for a CTO. We decided to meet up at a later date for dinner to discuss such options. For the purposes of this book, I will call him 'Danny'.

A few weeks later, though Danny was an hour late for our meeting in Shoreditch, we got into a restaurant, talked about ourselves, books we liked, and went through my company creation. We took notes and exchanged ideas. He was almost half my age, an engineer, but working as a teacher while also running his own solo 3D printing enterprise, which he wanted to scale up. I thought the partnership would work out well. We WhatsApp'd a few times more though he was busy with his teaching and business, sending me a demo of his 3D lithography prints. We parted ways for the Christmas holidays. But little did I know that that would be my last holiday abroad for a long while.

January 2020

I incorporated on the 20th. Why did I make it a limited company? Prestige. Ambition. Respectability. I wanted 3djacent to grow into a large company and not just be a sole-trader. There were to be multiple sections with marketing and sales, recycling, job pages, merchandising and affiliates, etc. I wanted to be the director/founder/creator of the 3D printing lifestyle management movement.

Later in the month, at the monthly entrepreneurs' get together, I met with my erstwhile business partner. We had a productive meeting following the first one at the restaurant and I felt confident. I got 'bling' silver metallic business cards and was ready to start creating marketing materials in t-shirts, mugs, pens, bags, calendars, mousepads, pamphlets, 'with compliments' slips, post-it pads, A5 note pads, headed stationery, and more. The whole exuberant overkill marketing process was in full swing.

Following that, over a WhatsApp meeting with Danny we started to brainstorm over merchandise, namely the tools and spare parts sets 3D printers required, including starter, intermediate, and advanced versions. We researched tools, spares sets, and tool boxes on Amazon, Alibaba, and other tools sites. We discussed how they would be bought and marketed, such as whether we would create our own tool sets or be affiliate-minded, or be a reseller?

We also discussed Facebook and LinkedIn ads to see how they worked and how we could use them. We checked out the Intellectual Property Office website to determine if and how we could protect our company names and ideas. There was a lot of work to do and I also had in mind another entrepreneur we had met who was a website builder and marketing guru.

February 2020

In this monthly entrepreneur meeting just as the group became known as Founder Nation, Danny was out again, but he seemed like a young puppy wanting to chat with everyone else rather than focus on the company. Alarm bells should have started ringing then. It was rather infuriating as I had brought him a few books on start-ups and a Formlabs eBook to read and research. Even though he was a teacher, he didn't seem to be much of a reader. The younger generation wanted things online.

But we persisted. Later in the month, I was invited to his flat in South London for dinner with his uncle and friend who were also entrepreneurs. I took over some of my newly delivered marketing goodies which went down well. We discussed more aspects of the business and what we could achieve.

Communicating was hard at times as he was younger, more socially active with his friends and family, overworked at school, and keeping his business afloat. Our conversations would be in the late evenings when I was trying to relax and I actually asked him to either message me before 11pm and after 6am as his text or voice messages would wake me up. Other times, he would be down the rabbit hole again studying Instagram and Facebook or a task manager like Mondays. com to see how they worked with ads and businesses. I felt that sometimes the details he obsessed over undermined the building of the company.

March 2020

And then the Covid-19 pandemic struck the UK. Lockdown. No more in-person meetings.

Danny and I spoke about this and he sent me a link to a news story regarding Covid-19 3D-printed medical valves for reanimation devices. His idea was to mobilize a small group of 3D printer owners and find out what the health service required and what they were struggling with in terms of parts. He knew it would be a big opportunity to help and bring some good exposure to 3D printing. Unfortunately, we could not implement this as stories started appearing online where 3D printers were threatened with lawsuits as their printing fibers for medical injectors and valves were not quality control cleared. There were discrepancies about whether this was fake news, the first of many over the Covid-19 period. So, we decided not to venture forth on the plan. But later, another successful plan would come in April.

The IoT conference had been due to start in March, but was postponed until further notice, so I was at a loose end. Hunting the internet, I found an STL file on how to print a hard-shell mask with a removable filter insert. It took 4 hours to print. Danny and I discussed the virus lingering on metal and plastic for hours, but I had read that alcohol rub would disinfect the mask so I could clean it and replace the filter frequently. I used a piece of cut-out hoover bag wrapped in a cutting from a pillowcase as a filter (as there were fears one could breathe in the fibers from the hoover bag) and tied elastic straps to the mask. While Danny and I also wondered about antibacterial coatings such as copper, I just used white PLA, later printing another mask in gray PLA. I wore it to work. It made me look like Batman's foe, Bane, and you could barely hear me speak through it, but it worked, and I loved it.

April 2020

During the lockdown, I was still working at my day job as a building manager, sometimes from home and other times in the office. But the ongoing lockdown gave me opportunities to do things I hadn't had time to do before or offer new experiences.

The first and best thing I did with 3djacent was to use my printer to combat the virus and keep people safe. I had read a BBC online article in March which ended in a link for volunteers wanting to 3D print faceshield parts for medical and emergency staff. The organization for this was the COVID-19 3D CrowdUK volunteer group. We would print the parts and they would be collected and delivered within the local region to medical personnel.

The downloaded file we used from 3D CrowdUK was for a Prusa RC3 model and we could use PETG or PLA. I tried both but preferred PLA as my printer seemed to print better with it. There were also other sterilization processes to adhere to with volunteers having to wear a mask and disinfect the printer and parts. This was definitely when a 3D print station would have come in handy. Printing so much after work and on weekends I also had to replace the bed mat and nozzle, so I bought a little cleaning kit and micro keyboard hoover. In all, during April, I printed 40 faceshield parts (upper and lower frames) which were then cleaned up, packaged in large ziploc sandwich bags, and labeled all under the auspices of 3djacent. I was very proud to do so.

May 2020

I completed my volunteer stint with another 20 faceshield parts. At the time, my other passion took over and that was writing books. And if you like or dislike this book, you can blame entrepreneur/author Daniel Priestly, either way.

I had been lucky to acquire his book *Key Person of Influence* at one of the entrepreneur meetings. There were lots of kernels of information, but the one that stood out to me as a writer was that to stand out in the 3D printing industry (or any other industry) was to write a book about it. To become an expert, or an idea catalyst, or a marketer. And since I had a different voice to add to the world of 3D printing, I decided that whether my company took off or not, I would still have an avenue to advance my 3D printing lifestyle and cultural vision. I'm not a 3D printing expert, but an amateur looking ahead to the future and all the adjacent possibilities for 3D printing. I had noted in my journal, I expected the book to be twenty to thirty thousand words long.

And even while I researched and collated my thoughts on this book, I was continuing to write my sci-fi books and also putting together a couple of nonfiction works based on essays and articles I had written years before. It was like the lockdown had released my suppressed creativity.

August 2020

I spoke with Danny about other volunteer printing projects including another project to print children's prosthetics, but I never got around to it. I told him about my plans to start writing a 3D printing book while trying to focus on the workstation and recycling aspects of the business.

October 2020

My daytime job and writing the other books during the summer and early autumn had kept me from really expanding 3djacent. I had even enrolled in an expensive copywriting course to expand my writing job prospects but didn't finish it (thus losing a good chunk of money – not a great financial decision). I had to stay committed to the company. By then, I had decided to stop using the GoDaddy site. It didn't have the range I wanted to convey my vision.

January 2021

Mid-January, Danny and I connected again and created a WhatsApp group between us dedicated to the creation of the 3D printing workstation. We went over designs, looked at possible competition, and I would send a short question on social media groups to gauge their views on such equipment, as per the below:

> Hi everyone, I've had an Ender printer for a couple of years now but never quite found the right surface or table to put it on. What do you guys recommend and what do you look for in a platform? Thanks.
>
> #3dprinting #design #innovation #additivemanufacturing

While my post got over 2000 views across social media sites, a couple of interesting posts stated they used tiled tables, tiles on any surface, or even paving stones on foam pads. The user of the tiles explained he could also use a white marker to write on them instead of on pieces of paper. I just thought why are they making do with make-shift surfaces?

Any craftsman would use a proper tool and equipment for their craft even if it's just the platform. Yes, it may be cheaper and more readily available, but to me, it should be the proper tool for the proper job, including the work surface.

Others sent in pictures of coffee tables, home-built benches, and racks whether in their shed, garage, or living room. One even had his printer on a butcher's block. Here were potential customers serious enough to want proper 3D printing furniture to carry out their craft. But there was nothing out there for them. I was crying out for the capability to be able to manufacture this for them.

Danny noted these down and came up with a small to-do list:

1. Contact table guy I know. [He knew a crafter of metal shelving who might be able to adapt his skills to what we wanted.]
2. Definitely [get] more feedback [from social media].
3. Study market, see [what the] pricing is like and what consumers are willing to buy?
4. Do initial sketch design.
5. Features to consider i.e. table top moves up and down and also rotates.
6. Add like [sic] feature bar to move product.

It was a good start and I hope we would kick on from there.

February 2021
I also paid the ICO (Information Commissioner's Office) their GDPR (General Data Protection Regulation) fee as I would potentially be handling customer data. There was always some kind of cost or company documentation involved whether confirmation statements, corporation tax, or annual accounts to file. It was one aspect of the business I didn't like as sometimes there were no reminders for payment dates.

I was also working on an idea for 3D printing-inspired t-shirts. I had ideas for using Merch by Amazon and their print-on-demand business as a good way to spread the 3D printing culture. As a proof-of-concept, I designed and created a couple of t-shirts through the company Vistaprint to see what they looked like. I had ideas for Craiglist as well and possibly re-selling printers there. I discussed these with Danny, who also liked the t-shirts I had created, but he was quite busy with teaching and then was on a half-term holiday after. Soon after, he was starting a university course for his teacher training. We had dedicated another WhatsApp group just for workstation design specs so I dropped in files for the t-shirts for him to view later. It took a while for him to respond, which was frustrating and nothing happened for a few months.

Near the end of the month, I had signed up to and started building a new company website on Wix, migrating away from GoDaddy. I changed the domain name and acquired new email addresses. I spent the next couple of months working on it with menus, banners, and site pages.

May 2021

Finally, I got a business bank account. I decided to use the online banking platform, Tide, as I wanted to have separate banks and accounts from my personal account. I found it a bit of a chore to link this new account to platforms like PayPal, eBay, Amazon, and cryptocurrency sites, but I had to make sure all my business went through this account.

In the meantime, I had been drawing up t-shirt designs and hiring a designer on Upwork to create a finished product which I would then use to create a t-shirt line on the Printful merchandise site.

I was also starting to think about the website's future on Wix. I spoke with several Wix designers trying to gauge if the site could handle the kind of expansion I wanted.

And with all of this going on, I had chosen to resign from my day job. I had decided to take the final plunge and go into business for myself.

I also felt I was falling behind on the business setup, especially without feedback from Danny. So, at the end of the month, I had another meeting with Danny at an east London pub. Even though by this time I'd had my two Covid-19 vaccine jabs, we had to observe Covid-19 protocols, order through an App, and wear a mask inside if not eating/drinking, so we decided to sit out in the beer garden. We discussed the business and the 3D print station idea which I hoped we would thrash out decisively. I showed him what I had done with the Wix site and he wasn't impressed with the layout or colors, stating I could get a better bespoke website for cheaper from a freelancer. As his thoughts on the website chimed with mine, I decided to go back to Upwork and Freelancer to find such a developer. My only concern at the time was that he would get bogged down in the minutiae, while I just wanted to finish one project and attend to the details later.

June 2021

I now had eight weeks of my day job officially left. After that, I would be self-employed. I was excited, scared, positive, and somewhat stress-free and stressed at the same time. I wondered how I would survive.

Even though I was working on what I called the 3djacent website v.3 (my design structure is shown below), I knew it would be too complex for Wix, so I had to step up my search for a freelance developer. I had asked if any of the Wix developers could accomplish this for me or recommend someone, but of those who could, their prices were beyond what I could afford. The new site was going to be quite large, with multiple services pages and interactivity, and the Wix designers could not accommodate my ambitions. I even asked my brother who is a website designer who recommended I look for a developer, not a designer, for a bespoke solution for the right cost but worth the investment.

Also, at this time, I had to think about my finances after quitting regular work. I had looked into various loans and grant schemes and thought I'd go for the British Business Bank seeking £10,000. However, without a regular income, I worried about repayments and what to borrow against as the company was not up and running.

But I was still coming up with ideas as expanded upon in this book regarding heat sensor generator-run 3D printing, the 3D printing XPRIZE, and transport cases.

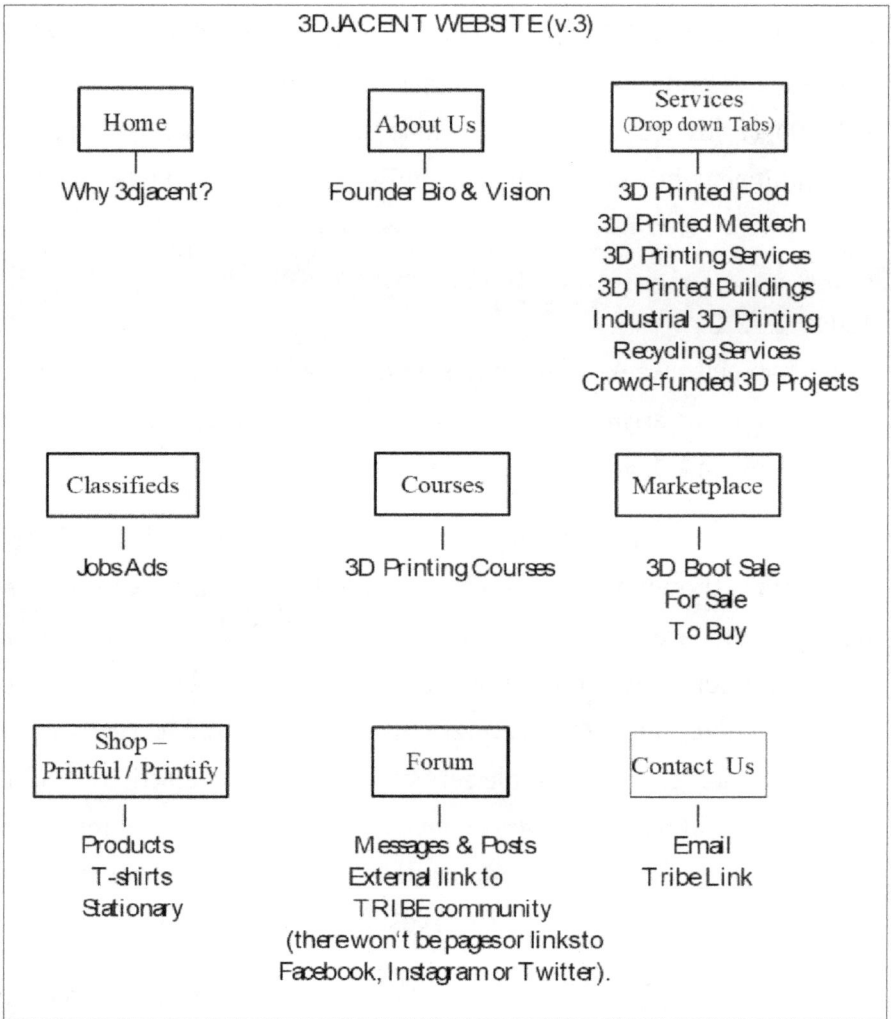

3DJACENT WEBSITE (v.3)

Home	About Us	Services (Drop down Tabs)
Why 3djacent?	Founder Bio & Vision	3D Printed Food
		3D Printed Medtech
		3D Printing Services
		3D Printed Buildings
		Industrial 3D Printing
		Recycling Services
		Crowd-funded 3D Projects

Classifieds	Courses	Marketplace
Jobs Ads	3D Printing Courses	3D Boot Sale
		For Sale
		To Buy

Shop – Printful / Printify	Forum	Contact Us
Products	Messages & Posts	Email
T-shirts	External link to	Tribe Link
Stationary	TRIBE community	
	(there won't be pages or links to	
	Facebook, Instagram or Twitter).	

July 2021

As the end of my work drew to a close, I knew it would be hard but was still confident I would succeed. Knowing I would be without money for a while, I had started to apply to the British Business Bank, but as I didn't yet have a viable business plan which was part of the application process, I decided to take out an unsecured loan for £7,500.00 (repayable over 5 years) for living expenses and to grow the company. I hoped that money would survive until October.

I decided to put the website build job out on Upwork for talented web designers to apply for. Out of about 70 applicants, I chose the best bidder, a team from Pakistan, who had experience in building websites for tech companies and quoted a reasonable cost. This was the Upwork and Freelancer ad that I put out:

Website design for 3D printing company.

Hi, I'm looking for a non-Wix or WordPress site developer, so a site from scratch. I have a current half-finished site on Wix, but I've realized that all the features I want won't be accomplished through a hosted site, hence the need for a site from scratch.

If you're familiar with 3D printing then that's a bonus. If the site is successful, I may look for a permanent website marketer and developer to join the team.

The £1000 cost is a holding placement until the successful applicant is chosen and we discuss the works. The site will be around 25-30 pages with affiliate links and GoogleSense Ads. I will supply most of the text and a mock-up of each page, though images will have to be free or low-royalty fee.
Some of the site pages are included below with more to follow:

- Home Page—It's a simple page with a site logo, banner, taglines, and menu.
- About Us Page.
- Services pages—Drop down tabs for 6-8 services. At the moment the affiliate company sections on my mock-up are placeholders to give an idea of design.

- Marketplace—for affiliate and customer buy/sell opportunities.
- Shop—I have a shop for company-branded items. I'm currently using Printful as the product supplier.
- Forum—Members can sign in for discussions and comments.
- Contact Us Page.

Thanks and I look forward to seeing your applications.

My briefing for Webefy Today, whom I chose for the work, included an update on my website v.3 idea:

We are a 3D printing lifestyle and management company with services and products adjacent to 3D printing. Our demographic will be all ages, experience, and income in 3D printing. This will include corporations, small businesses, solo-preneurs, and students. Those looking for products, services, jobs, courses, and networking on a forum.

Most of the sites are corporate catering for corporate customers. 3djacent will also cater to small companies and indys. We will have a forum for networking, a branded product shop, and new products to help with a lifestyle in 3D printing.

Mock-ups I have of page examples I will be using are:

Home | About | Services (with 8 drop-down tabs with differing industries like 3d printed food, buildings, medtech, services, industrial, recycling, Indys, crowd-funded) | Classifieds | Courses | Marketplace (Affiliate sales) | Shop (3djacent branded merchandise) | Social Media (with Blog – Surveys – Forum -Pictures – Events- Stocks – Discord – YouTube pages) | Contact Us.

[Author note: The Social media section was later expanded:]

There are 7 drop downs here. Blog, Forum (with external link), Surveys (with external link), Wall of 3doration (pics - with reference site link), Discord (with external link), YouTube (with external link), and contact us.

[Author note: The 3D printing stocks page was eventually discarded]

Members can add videos and pictures to social pages. It is free, just need name, email/contact.

Merchandise: T-shirts (Over 30 designs including signature Spoolhenge collection), office and stationary wares, currently on Wix site. There are 2-3 hardware products yet to be designed and sold.

In the end, I paid $1500.00 (£1200.00 including bonuses) for the website and the work would take 8-9 weeks to complete due to the complexity of the site and their other ongoing projects. I was assigned a dedicated project manager who would interpret my needs as I would be supplying drawings/mock-ups of the pages with text to introduce each section and tab I wanted. There was a lot going on in July, with work ending and having to start designing the new bespoke website.

Next, I wanted to tackle the 3D printer travel case design. Searching for manufacturers I had flirted with the idea of using the Wikifactory which catered for collaborative manufacturing, but as mentioned in Chapter Four, I had contacted Flightcase Warehouse. They quoted me £245.00 + VAT for the prototype. They seemed a simpler option and I was happy to pay that, plus if the case worked out, I would resell them with a healthy markup. I had recently bought a 3djacent branded 1.2m roller banner as a backdrop display and a telescopic mobile tripod so I could take pictures/videos and start creating ads for my site and my YouTube channel. [Author note—seriously, the telescopic tripod with remote for taking pics was a godsend. Highly recommended for your videos and pictures.]

Around this time, the Webefy project manager, Muhammad, sent me mock-ups for the homepage. I wasn't too impressed, but it was a start. But he got back to me three days later and they were much better, though there was more to do. I had to veer a course between not wanting a fully corporate-looking home page, but still making it look welcoming and professional.

I continued to map out the website, drawing the menus and drop-down tabs, and started sorting the 3djacent documents folder on my computer as it had become messy with duplicate entries and misfiled names. I do like a tidy documents folder, easy to find info, especially when it came to filing taxes, invoices, and account details. I needed to start keeping better track of my purchases, ideas, and progress. A Top tip for budding entrepreneurs.

One thing I did not get was VAT registration. My application was rejected. I only wanted it for better business authentication, but as I would not be earning the threshold amount ideally required for VAT registration, I didn't apply again.

As I completed the final days of my day job, I was looking forward to events I had booked, a potential re-selling opportunity I had come across on LinkedIn, and I had come up with the 3D printing game idea. And as I knew I'd be meeting Danny, I put together so-called 'Board meeting' notes to try and steer a clear direction.

'Board Meeting' Agenda and Notes
29/07/21

This is for us to discuss and expand on. I've been reading a lot (the books in #6) and they've given me big ideas so we can reach out more and connect with more exponential technologies and companies.

I'm positioning us as a 3D printing lifestyle management company incorporating consultancy (with our directories), retail (with the marketplace and shop) and social (with jobs, forum and blog).

1. Website update – 'About Us' and 'Blog' pages. I had sent over examples for you. I've asked them if blog page would be for Vlogging as well. Be great for members to connect and advertise that way.

    I'll also be looking up affiliate sites for us to join so we can start hunting advertisers and clients. In one of the books below one of the founders of a large company stated 'if you're not embarrassed by the first website or product then you've launched too late'. So while we're waiting for the new website, I'll go back to the Wix one and put up a notice stating we're open for business while our new site is under development. That way we can launch and get some funding/sales in. Hopefully, we can then transfer over the affiliates and members to new site.

2. Services and Suppliers database – I started a database of companies, groups, and suppliers to contact, plus the products we currently have. There's so much to add as I'll also be looking through the directories of other companies to add to ours. Do you think we need anything else?

3. 3D Print travel station prototype: order £312 – I've bought the prototype for the travel station and should get it by the end of August. It can fit an ender printer, spools, tools, and a shelf. Hopefully, the top can be used for a work surface. They should be able to make it

in different sizes for other printers and add other features. If it works out, I might take it to conferences to show or make a video for our site, YouTube, etc.

4. eSun insoles as reseller. As you saw from LinkedIn video this looks like a great product. She does keep contacting me to see if we want to be a distributor or partner. I would look to be a reseller finding clients for her here (as they're in China) or maybe even buy one and use ourselves to make insoles. Questions I have to ask her are: 1. cost of the equipment and materials 2. delivery schedules, 3. demonstration, 4. longevity of insoles, 5. contract issues.

5. Caribbean partner – From work, one of the contractors stated his friend is a tech designer and would very much like to know more about 3D printing and what we are doing. There is a lot of money in the Caribbean Islands with links to the US so having an affiliate or partner out there would be a bonus.

6. Reading materials – I will send you these 2 short eBooks – 1. The State of 3D Printing (potential client targets) and 2. Wix ecommerce handbook, 3. the three books by Peter Diamandis and Steven Kotler – ABUNDANCE, BOLD, and THE FUTURE IS FASTER THAN YOU THINK. These books gave me ideas on exponentials and the business. I signed up to Diamandis website and get free newsletters from him (basically summaries from his books) but well worth it. With ideas from these, I'm trying to connect 3D printing with cryptos, gaming, and global warming solutions.

7. 3D print game idea – I think we can really make a lot of money from this and also educate people about 3D printing. It's called 3D

Printectors and basically when aliens invade Earth, the only way to defend Earth is to 3D print weapons and equipment in a secret bunker. You have to set the printer design, select print modes and wait for the print all while shooting aliens. It's a Multi-role player, first-person shooter game as you protect the 3D printer. Members can play a free version, but ultimately have to subscribe for a fee. We could then sell it as an app.

Making the game will be expensive, but as there's no such 3D printing game, I'm hoping a game designer would take a nominal fee and wait for sales to come in and residual payments. What do you think?

8. Merch – I think I have a good stock of T-shirts. What do you think of the designs? Do we need hooded tops and what other fashion items do we need? The office stuff I got from Zazzle, but we can't resell items they had created by others so I'm looking for notebooks, notepads, post-it notes, with compliments slips, postcards and other items elsewhere to resell. I should be able to find reseller sites for those.

9. Domains - Should I buy back the original .com and .co.uk domains from GoDaddy £89.99 each. Or keep the current .com and .co.uk ones?

10. Events
LinkedIn – Exponentials Club August 3rd 6pm – 7pm (online)
IoT Tech Conference - Sept 6th-7th at the Business Design Centre by Angel (I paid for a ticket last year but it was postponed until now). There's a virtual event from 13th to 15th as well.
London Tech Week Sept 20th-24th - venue TBC. I've signed up for this as well (free)

Looking back at my notes and WhatsApp messages, Danny had phone issues and kept stating he would read through my messages and give me feedback on the Board Meeting notes, but he never got around to any substantial input and kept saying we'll meet and talk about it. This was quite disappointing to say the least.

Around the same time, I was still liaising with the website developers over designs. I had also designed and bought a 3djacent 'staff' short-sleeve polo t-shirt and a cotton button-up shirt embossed with company logo for events. Plus, it would make a great shot for the 'About Us' page.

I was still ambivalent about Danny being a business partner/co-founder. Even as I pressed him for a short bio and whether he wanted a staff shirt or even for feedback on the new website designs or on the reselling/affiliate opportunities he seemed reticent. His teaching role and own 3D printing enterprise were his main concerns and I seemed to be the distraction. We had discussed the site with some constructive criticism, but it wasn't enough. Further journal notes at the time stated that I needed more consistent meetings and ideas from him. So, I carried on myself and signed up for affiliate sites Awin and later, Trade Doubler.

August 2021

During the month, I finally had a WhatsApp video meeting with Danny. However, the meeting wasn't as deep as I wanted nor as positive thinking from him.

However, the website guys had produced some great directory pages for me. I was happy on that front. We had multiple Skype calls and could upload images and updates on what I wanted. This was also the time my idea on the 3D printing Recycling Lab was starting, the beginnings of the collaborative recycling company, which added another page to the website under the Recycling category tab. My drawn-up pages for Classifieds, Marketplace, and the several Social Media pages were also sent out. I wanted to be specific on what was required and what could be achieved on screen by the guys. I learned a lot from them and how what I drew and wrote in their various frames would present themselves on screen.

Mid-month, I had a couple of meetings, one of which I was excited about.

A Chinese company called iSun/eSun was developing 3D printed insoles. For their process, each of the customer's feet were scanned in a bowl-like device and customized insoles could be printed in minutes. It was a great business opportunity. I had a Zoom call with one of their representatives and a video tour of their studio. I was impressed with their other products. When I discussed this with Danny his reaction was 'Deffo selling and distribution will be sick.' But there was a setback. I wanted to be an affiliate for them marketing and selling their product for a commission, but they wanted resellers working through their 'cooperation modes' who would be trained up on the device, buy the equipment totally over USD 10000, stock them, and be the demonstrator/distributor. Well, I just didn't have the money to be buying anything like that. During the month, in back-and-forth messages, we couldn't resolve the impasse so I had to pass on the opportunity.

The same day, I had a video meeting with an Amazon rep to set up my Amazon business seller account. I was successful in jumping through their hoops and now I had to link and upload the products from my Printful shop.

As much as I kind of hate marketing on social media, I had to reboot my 3djacent Facebook and LinkedIn pages and gear it towards business networking. And if I found that difficult, then I found trying to link my Printful stock to Amazon a nightmare. I spent long hours trying to authorize EU/UK and US marketplaces, filling out forms, syncing products, creating billing preferences, etc. There was no real joy in this or much help from the onsite notes. It just seems companies like their gobbledygook when just plain-speaking and maybe annotations to screen diagrams would do. So, I left that to stew for a while. Unfortunately, a few days later the syncs and uploads just didn't work and nothing seemed to help. I decided the Amazon business seller wasn't for me and I had to try something else.

I had also signed up for a Daniel Priestly Zoom webinar on lead magnets, which I hoped would give me more insight into gaining customers. I had also ordered a book on affiliate marketing and an eBook on Community Gamification. The latter was to help in understanding how to engage customers with competitions, games, and rewards.

The affiliate marketing book by Dean Holland was a 'free' book only costing the 'inflated' shipping and handling costs, a so-called 'freemium' product designed to not just introduce the entrepreneur, but also to create a passive revenue stream. It was a clever scheme, but what I didn't like was the quasi-pyramid scheme attached with it. [Affiliate marketing is legal, but the hard sell tactics and my gut feeling turned me off somewhat]. As the ad I had signed up to had been on Facebook, I started getting tons of emails to help this entrepreneur spread his affiliate marketing business while herding them toward an expensive course at the back-end of the business which made the real money. It's a growing trend and one I was slipping into. I just needed the back-end and freemium products for 3djacent.

The Daniel Priestly webinar was less a general lead magnet talk than a launch of his ScoreApp funnel marketing service. It was basically a survey/quiz which was set up on your website or social media pages to entice 'warm' customers in and to help customize your business through detailed feedback. We participants got a month to test it for free then it would cost slightly more than other services with the same process, depending on the different templates and pricing plans. I had to figure out how it fit my business model. I signed up, picked a template, edited and sent out my first survey to 3D printing-related groups on LinkedIn and Facebook and waited for any leads.

In a busy August, looking to possibly raise money and learn about trading, I had invested money on eToro. They provided live ninety-minute courses (or saved to their online academy) on how to invest safely and what was good to trade in. I learned a lot and had spread my money around in technology and natural resources. I set limits and cashed out when appropriate, in most cases, while other times I was too stubborn and lost more than I wanted, which was ideally less than fifty percent of my investment. And I was also invested in Coinbase. Top tip, for 'free' cryptocurrency do the quizzes on new coins then you can convert it to any other crypto or even cash when you have enough. It helped me out in later months when I was strapped for cash.

September 2021

I had some fun at the beginning of the month. I had started crafting scripts for videos I would be filming for the website and YouTube as a way of introducing myself and the company. Needless to say, doing it yourself as an amateur can be a bit of a learning curve. So, off I went to my local park in my staff t-shirt, mobile phone and tripod, and my script.

It took me about two and a half hours to film a decent 3-minute video, through several versions, flubs, windy takes, cloud/sun changes, people in the background, and noise from the park staff mowing the fields, etc. I had the scripts printed up, glued to cardboard, and wedged in between the mounted mobile and the tripod frame; that way I could read out the script while walking. The tripod came with a remote so I could start/stop the filming with a press of a button. I filmed the 'About Us' intro and what 3djacent was about then filmed the 'Recycling Lab' promo. I uploaded it to YouTube and sent it to the website builders. Feedback from them was that my walking around in the park holding a slightly wobbly tripod was amateurish, but I actually thought it was more natural and welcoming rather than sitting behind a desk or in an office setting. I enjoyed the experience.

Later, in the month, as detailed in the chapters above, I attended the IoT Expo on 6th and 7th of September, which opened my eyes to the wider technological ecosystem and its potential with 3D printing. I was slightly miffed though as I had 2 sets of business cards on the way; one was another batch of the silver metallic cards complete with a snazzy card holder and the other was black with red writing, the colors of 3djacent. I wasn't happy about the late deliveries. But while there, I came across a BBC online article about the death of physical business cards, especially in a pandemic world where tactile objects would be digitalized for better access and safety. Straight away, I looked up digital cards and came to Blinq. I signed up, created my cards with a handy QR code and used it a few times at the IoT conference. Sometimes, coincidence and circumstances beyond your control can actually lead to a lucky break.

After the second day at the conference, I attended the entrepreneurs' MeetUp in a pub not far from St. Paul's. One piece of advice given to me by SEO marketing manager, I'll call Sean, was to hire other people to carry out tasks I had no expertise in or needed help with. One such task was my upcoming corporation tax and accounts, which I had no clue about. So, after talking to a freelance accountant at the Meetup, and after a Zoom chat with her, setting up my Xero account which made it easier to track my invoices, and paying her £400 fee, she was able to sort out the accounts and tax for Companies House and HMRC. It wasn't pretty reading.

Up until then, I had spent almost £3000.00 setting up the company and on marketing materials but returned less than £100 in sales, most of which was from the 3D printing of the faceshield parts. But, at the same time, the process made me more aware of my spending and accounting processes, so I made sure I had filed my hardcopy invoices, digital copies, backed up on laptop files, and used Xero for automatic accounting.

Another choice I had made while speaking with others at the Meetup was to start up a Shopify shop to replace Amazon. It seemed less finicky to set up and

I could link it to Printful much more easily. I was still creating t-shirt designs and placing them on Printful, but needed a graphic designer to complete other designs. So, I turned back to Upwork for that. About a week later, I had imported and synced my Printful products to Shopify. So much easier and better than Amazon, so I canceled my business account with the latter. And I had a brand new product to add, too.

Remember the 3D printing travel station I had created and ordered from Flightcase Warehouse? Well, that arrived all tightly wrapped up and I had to drag it upstairs to my flat. It was bigger than I thought it would be and a moveable shelf I wanted was fixed in place, but it made a good-sized compartment for spare parts and ancillaries. I took pictures of it inside and out, measured it, and prepared a script for a sales video. I practiced wheeling it around, handling the butterfly locks, and operating the telescopic handle. Now I wanted Flightcase Warehouse to build the 3D print station as well. But for the moment, there was only one thing missing for the travel station ad and that was the branded 102mm x 206mm roller banner I wanted for the background. That arrived at the end of the month.

From 20th-24th, I attended the virtual sessions of London Tech Week, which was an incredible experience with great speakers and such idea-spinning material, which I knew would be crucial for advancing the 3D printing world.

After trying out Daniel Priestly's ScoreCard App, I decided to ditch it as it led to no new leads and it was costly for my needs. I looked at Survey Monkey and Lime Survey choosing the latter since it was free and I could create more surveys.

And I had also come up with an idea for my 'freemium' product. I researched 3D printing log books on Amazon and came up with my own log book/ workbook/journal. The beginnings of an idea for a 3D printing TV show also took root.

October 2021

This was going to be my big month, basically 'launching' 3djacent officially at the Paris 3D Print Expo from 20th-21st.

But first, I had to deal with my website, as I felt the guys were about 2-3 weeks behind. We had another Skype meeting and I had to tell them to sort the errors and pages out. I wasn't happy and also gave them a deadline to finish work the next week. I needed to start earning from the site.

I had invested more money for the conference in goody bags with branded QR-coded A5 flyers, pens, business cards, bookmarks, and sweets/chocolates (an important networking ingredient for hungry exhibitors on the floor, I had learned from my time at the entrepreneurs' Meetups). I had booked my transport on Eurostar, the hotel was less than a 10-minute walk from the venue, and now I just needed a working website.

After another meeting with Webefy and going through the site page by page, I was happier. They were confident of meeting my deadline. A third meeting within the week and they seemed to be on course. I was basically going through each page making sure links, features, and videos worked; that my text and pictures were aligned, and that members could sign in, add their material, leave feedback, pictures, and enjoy themselves on the site. I wanted to build that community.

I had hired an Upwork freelancer to finish up on t-shirt designs and on removing the living room background from my travel station photos then uploaded them to Printful and Shopify. My accountant had also completed the corporation tax and accounts, so that was a big relief.

And if they weren't distractions enough, I had started writing this book about the then-named *Future of the 3D Printing Culture and Community*. I also spent three hours filming two ads in my living room with the banner background;

another intro and the 3D printing travel station ad, which I then uploaded to YouTube and then readied it for the website. Using the banner, speaking and using props, and trying to sell myself and products was another good learning experience. Lastly, there was another entrepreneurs' meeting, where I had met with a possible supplier for 5-gallon plastic water bottles that I could potentially recycle into filament. I kept that on the backburner as another problem was developing.

Somehow, after all the positive meetings, the website guys just weren't delivering. I was quite angry with them as they were still getting details wrong with my directories (where companies could upload their information). They still had to fully complete the home page with a call-to-action line, the events page, and get the social media pages working, especially the Discord and stocks page links. The main problem was now a site host. I had started with GoDaddy, went to Wix, and now the developers wanted me to go back to GoDaddy. It was two days before the conference with more money being shelled out on the website hosting, but I had no choice. So, I started a new GoDaddy account with a new domain name and emails, and waited for the site to go live.

Paris 2021

I had been a couple of times before to visit my then girlfriend/ex-fiancée (who accompanied me on my second day at the 3D printing conference as my translator), so I actually found my way around easily though my French was still relatively non-existent.

As detailed in Chapter Eight, on the days, goody bags in hand, I scoured the conference center networking and trying to gain contacts and see amazing 3D printing technology at work. But guess what? My website did not work. I got a call from Wix while I was at the conference to resolve the domain transfer to GoDaddy. So masked up and tucking into a corner for privacy it took about twenty-five minutes to sort and for the Wix rep to email instructions on how to point my domain to GoDaddy and then I had to let my web builders know, which took them a further nine hours to resolve. It was a nightmare, but finally, I could say I had launched 3djacent Solutions Ltd in Paris.

At the same time as successfully overcoming any social anxieties and navigating the Covid-19 restrictions at the conference, I also began to learn about the limitations I was already going to encounter when trying to set up a company with no real sales strategy and no products to sell. I was generally asking the companies there to sign up to my company for free, join a nebulous 3D printing community, and I would somehow market them. Maybe I'd offer affiliate services, a place for them to hang out on the social media pages, and tout their services, products, courses, and jobs. I wanted to be the go-to 3D printing guy, but beyond that, there was no clear plan.

I'd fallen into the business trap of creating a service that had no problem to solve or a customer base. Yet!

And it didn't get any better when I returned to London and followed up on all the contacts I had made. Most wanted reselling opportunities and others didn't respond. It was heartbreaking after all the work I had put in. But there was a silver lining. My Parisian friend wanted a little statuette printed for her in bronze filament, which I did, and she became my first paying customer (at a hugely discounted cost), for a service 3djacent didn't formally provide.

A few days after getting back, I met with Danny in a pub. We went over my Paris visit, talked about my recycling idea, and he nitpicked over my website, which I wasn't going to change for him. It was becoming more frustrating to meet with him as nothing constructive was coming from him. It wasn't the website that was the issue, but my business plan I needed help with.

I had set up three meetings for early November, one with a 3D printing site I had seen on Facebook, the other with Sean and his website media marketing business, and the last with an entrepreneur looking for businesses for Tokyo.

November 2021

Late October, I had followed a link on Facebook and watched a video by the Institute of 3D Printing (Io3DP). The director, Ed Tyson, was very knowledgeable and helpful, discussing ways of making money not just by printing and selling the same thing as everyone else, but by choosing a niche and printing simpler cheaper products with a higher mark-up. It made sense and was an attractive business model. I made an appointment to speak to Ed.

When I had my scheduled call with Ed, he asked me what I wanted from the 3D printing business. I think I confused him with all my ideas for my company and wanting to partially build it through recycling initiatives, affiliates, or reselling. He sent me more info with details on what type of products to print and sell, like replacement hood ornaments, drone parts, or webcam covers. But in truth, I think I was too scared or insecure about my printing abilities and making products myself for some reason. Maybe I was burnt out on the idea of actually printing and wanted to follow other adjacent ideas. But what could I do? So, after a few test prints of ornaments and webcam covers, I decided not to pursue that avenue.

My next call was with Sean the media marketer. Sean and I had always got on at the Meetups and he was always trying to get me to take his offer of SEO and marketing assistance. Not a hard sell, but politely persistent. Our call was informative, even if some ideas were against my intuition. His feedback after looking through my site was that it was very broad and overwhelming with lots of offers which could confuse customers. He wanted to keep it simple and go with a product offer with B2C funnel hacking and creating two communities: one with products to the 3D printing community and a 3D printing community platform. He liked the niche product idea from the Io3DP and wanted me to come up with a selection, though I was already cool on the idea. Once that was set up we would then look at paid ads and platform building.

Two products I was creating were the 3D printing workbooks; the 25 and 50 projects journal/organizers. For these, I had completed the pages, designed my own covers (black with red lettering to fit the 3djacent brand), hired an Upwork freelancer to format the correct cover size for the books which were to be hardbacks, to my delight (as I felt workbooks should always be hardbacks for durability), and then uploaded to Amazon KDP. However, they would not be available under Prime orders and strangely, the books printed through Amazon were in the US A4 size which was slightly stouter than UK A4 size. Nevertheless, they would be 3djacent commercial products offered on the website as well. Secondly, I was trying to increase visibility by investing in two 3D printing Kickstarter campaigns and by contacting a few 3D printing companies on Facebook to sound out my ideas.

The last meeting I had set up was through a LinkedIn Access to Tokyo seminar (A2T - Industry 4.0 Webinar). I wanted to follow up with Adam, one of the speakers to see what opportunities there could be in Tokyo. The meeting didn't last long as the offer was really for having a physical presence in Tokyo, not just an online one, but he did send me more information should I reconsider in the future.

So, three meetings and only one real viable option and that was with Sean. We spoke again and we decided to pivot the company away from my sprawling all-in-one and to focus on the niche products and the branded shop. But there was one more loose end to tie up.

Unbelievably, the website was still not totally complete. A few areas and issues needed clarifying, the stocks page was now being deleted, and there was a mention of using Stripe for on-site payments, but I scrapped that idea as it would have cost another $500.00 to set up. And I was fast running out of money from the original £7500.00 loan. Once those issues were cleared and I was satisfied with the state of the website, I finally ended the contract with them three months later than anticipated.

Money well spent? For what I wanted, yes? They had delivered what I wanted. For what the business achieved? No. However, that wasn't their fault. It was my duty to push the site, but I wasn't up to the job and didn't know what I didn't know about marketing and customer base building despite reading what other businesses had done or speaking with other entrepreneurs at meetings. I think my problem was I wanted to be different and do things differently. I didn't see it as trying to get something for nothing or laziness, but I was just so tired and wary of the same old social media schtick. It never interested me. It may have worked successfully for others, but I was trying to do something organically. Which in the end wasn't going to work.

The next time I spoke with Sean, it was more in-depth about how I could start making a profit. We brainstormed on subjects and decided to ditch the niche products and pivot the company to creating 3D printing courses. He felt we could earn US $5000.00 to $10,000 per month with video online content with a price point for the course between $300 to $400. I thought that was too high, but he had ideas for selling the course to Middle Eastern customers who had the money to spend. It was what I needed, but I wasn't sure on how to set up the course. My course had been more practical with little written theory, so I needed to research it.

I had just about run out of my runway money. With no other recourse, I decided on taking out another loan. In not choosing the British Business Bank and in failing to get a loan from my bank and from my current loan lender as it was too early for an extension, I had to find another lender. I was able to secure a £5000.00 loan from another lender, (repayable over 5 years) and my plan was to be making money from the course in order to pay both loans back.

I was very happy when I got an affiliate/referrer chance with a company that manufactured 3D-space photography cameras that could produce digital twins, allowing for virtual tours of buildings and spaces, plus creating detailed floor plans. There was a free demo and customers could get started without experience. I thought it would be a great product and I would get a commission for sales.

I sent their tailored-message links and my adapted emails to my former employers, other property management companies, contractors and surveyors I had worked with, building developers, 3D printing groups on Facebook and LinkedIn, and advertised on my site's Marketplace page. It may have been the cost of the camera, but as my former manager stated, it seemed like a suped-up GoPro, which of course it wasn't. I got a good click rate, but no sales, which is what earned the commission. In hindsight, I possibly could have talked to Sean about this, but never did. He could have made all the difference at the time and saved the company.

In another setback, the self-publishing company I usually used for my sci-fi books now had a policy of not publishing workbooks and its ilk as they didn't find them of value, so I had to find another publisher besides just Amazon.

I met Sean at another of the entrepreneur meetings and we discussed setting up the course through his services. I told him I would need another loan to be able to afford his services and he said to look at it as an investment. His costs would come to £5000.00 paid in two installments for setting up the online infrastructure, creating an automated email/call system with a closer and virtual assistant to secure potential customers, and hosting the course on a new website. He would send a breakdown of the costs. I paid the first half of the fees from my new loan and he got his company to work. He would also use platforms I had not heard of before, like Zapier (a workflow automation tool) and Twilio (a customer engagement platform) as well as me creating 3djacent accounts for the course on Twitter, Instagram, Telegram, and Facebook, which I was reluctant to do as I was tired of social media as a whole, but needs must in business.

During that time, I attended another virtual event called Silicon Valley Comes to the UK (SVC2UK) from 15th to 18th. It was another informative conference as detailed in Chapter Eight. I loved attending these conferences during the year to get out, learn new things, and to be out of my comfort zone with learning about new technologies and terms I didn't know existed. But if these technologies were going to change the world, then I needed to know which could be allied with or adapted for 3D printing purposes and accelerate its adoption by more people. I hoped I was on the right track with 3djacent and pivoting to courses for the foreseeable future.

Sean and I had another conversation and we quickly saw there had been a miscommunication. Sean thought I would be teaching the courses. But I had to let him know that I had no expertise in running such courses. I may have taken a course and printed things, but that was different to running a course where qualifications would be on offer. So, I started researching courses and collating the best bits until I devised my own course as described in Chapter Seven. But I still needed someone to teach the course. Plans for the course included optional selling and dropshipping of 3D printers to the students, sourced from affiliates or resellers with costs recouped by the course costs. They would get a free workbook as well, based on the original logbook/journal/organizer template.

Away from the course and trying to drum up sales from the Shopify side of the business, I started a one-week Fiverr campaign for my t-shirts. The marketer, from Nigeria, had good stats so I tried her for a while, but I wasn't sure if I had the right campaign or what markets she was advertising to as there seemed to be no sales from the 725 or so clicks. Was I doing it right? Was I missing anything? I just wasn't getting a break anywhere.

When I talked to Sean again via Zoom, we went over the course and the need to outsource the teacher. We discussed what to call the organization that would teach the course and after some brainstorming, I came up with the Academy of 3D. We got the domain name and emails on GoDaddy for £21.00. Sean suggested setting up a Trustpilot account for future feedback and then we went over value stacking. The course would now be c.£145.00 rising to c£200.00+ if a 3D printer was required. I then sent out a message on the 3djacent Facebook group if any of the members would want to take the course and I got some very positive feedback. I was feeling elated. Things seemed to be taking off.

I started drawing up the course syllabus and feeling out qualified people who could give the course with several Facebook group members offering. They were either engineers, designers, or had 3D printing experience. I was also picking up several affiliates from Awin which would have been of help.

In the end, I talked with Danny, who seemed quite keen on undertaking the course. He had the 3D printing experience with an engineering background, and was currently teaching engineering to students. Sean had wanted the course delivered by December 6th, but Danny had a tight schedule with a university audit on his teaching course. His estimate was to have the course ready by mid-December, at the latest. He wanted to do it properly and could probably film the course using his school's resources. Plus, he wanted a company-branded shirt, so he looked professional and associated with the Academy of 3D. I obliged him and ordered a couple of white cotton short-sleeve shirts with a logo I had quickly devised for the shirt: ʌo3D. He sent me a sample of a combined video/PowerPoint presentation he had done before, which was the type of presentation I was going for. By the beginning of December, I was feeling happy.

December 2021

While I was waiting for the course to be produced, I used the same Upwork freelancer to format the covers for the course workbooks/organizers, which included the 25 and 50 projects books and the report/logbook. These 3 books would be A4 hardback books and have a blue cover with white lettering for the academy to differentiate from the black and red commercial 3djacent workbooks. They would be produced and sold only through Lulu Publishing.

I also then decided to publish the same books as A4 paperback (as hardback for these wasn't an available option) for 3djacent, so the covers had to be formatted again as the paperback and hardback books would have varying measurements due to the spine width. But at least they were the UK A4 size. The 3djacent workbooks were all uploaded to Amazon and to the 3djacent Shopify page. So, I now had 3 sets of workbooks covering the commercial and academic businesses. I just had to wait for sales.

Shortly after, the two white cotton shirts I had ordered for the course presentation arrived and on the following day, I created Twitter, Facebook, and Instagram accounts under the name 3DGOAT (Great Opportunity and Training). Now, Danny had to deliver. He drove by my place to collect his shirt, but his update on the course was not satisfactory. His school work was piling up and he was on a university course for his teaching role, so his time had been limited, especially with Christmas arriving soon.

It was not only Sean's DIY marketing website waiting, but also a large online learning platform. I had contacted them in order to upload the course for potentially thousands of their customers worldwide. It would have been a great secondary revenue stream for the course. I hoped to be able to sign with them in January. I just needed the course completed.

Danny then sent in a PowerPoint sample of module 1 of the course. It was okay, a little light on detail, and just under ten minutes, but it was a start. I asked him for more FDM details in the practical part and an activity or quiz as I wanted to give more value for money in the course. But once everything was together we could then produce the finer details. At the same time, I had contacted those interested from the 3djacent Facebook group about producing courses for 3D modeling with one member quite keen to do so.

By the end of December, I was at a low ebb, financially, but positive in that we had an out with the 3D printing course, a back end that would raise profits. I was getting a bit desperate now. There was no money coming in from any 3djacent revenue streams (t-shirts, books, travel station, office products, or affiliate efforts). And there was still no course.

I had already resorted to selling personal items like DVDs, books, and other things on eBay and Facebook. It was that which kept me going, buying food and paying the bills. I had to cash in all my crypto and eToro funds and max out my credit cards to survive January. I went to my bank for a loan or a new credit card, but no dice. I already had two relatively new loans and was already searching through over thirty lenders for another loan. My credit was shot. But, luckily, I got a small personal loan (repayable over 4 months) for £400.00 from a loan company. It wasn't a lot, but I would survive for longer until the course was complete. But, it meant I could not pay Sean for his marketing services. He understood my dilemma and we agreed to talk again after the holidays.

However, Danny was a different issue. We had discussed the idea for 3D printing courses to start at the beginning of December. But my decision to place my faith in Danny was proving disastrous. Even as I relayed to him the severity of my financial situation, I got excuse after excuse as to why the course was not completed. Days slipped into weeks until it was Christmas with nothing to show.

He finally told me that his delays were exacerbated by the fact he was suffering from depression due to an illness in his family and other circumstances. He couldn't focus on the course and his mind was not in a good place, so he had personal challenges to overcome. I felt both angry and betrayed. At the time I had no compassion for him. This was my life and business on hold and on the line. I had trusted this guy, left him time to get the course done and quite frankly thought he could raise his game and deliver the course despite what he was going through. Overcoming his challenges would have focused his attention away from his problems. But I conveyed my sorrow for his situation and hoped he would have the course ready after the holidays, which he hoped so, too.

THE STORY OF 3DJACENT SOLUTIONS

January 2022

Just after New Year's, Sean contacted me. He could not deliver the new Academy of 3D (Ao3D) course website and services until he was paid the next full invoice of £2500.00 for the DFY marketing system, even if I got the course uploaded. Fair enough, but I wondered how he would get the money without the course being set up and taking the payment from the profits.

I felt very let down by Danny as he was 5 weeks overdue and it was affecting me, Sean's DFY company, his team, and the e-learning platform I had lined up for the course. We needed to have a talk week commencing 3rd January as the platform needed the course uploaded so they could add it to their catalog for the year. Things were as desperate as ever. I was being placed in an impossible spot by a marketing professional and an unreliable friend.

Then I heard from Danny. He seemed to be feeling better and he said he could have the course by January 11th. I could then upload it onto the online learning platform and start earning enough to then pay back Sean. However, Danny was still coming up with excuses as to why he couldn't finish the course. It's like he thought I had all the money in the world waiting on him to finish or to pay someone else for other courses.

He finally delivered the video/PowerPoint module 1 of the now-titled 3D Printing Revolution—Beginner's Course. It was basically the same theory session as before, just over 10 minutes with a promised 10-minute practical lesson. He also stated he wanted to add some activities over Zoom or webcam but his equipment was not what he wanted to achieve this. I was disappointed as I could not sell what he had sent me or price the module by itself, at least on my website. I was reluctant to upload to the learning platform as well. I needed the whole course. I re-sent him the full syllabus outline to encourage him to complete the whole course. But there just seemed to be a continued sulky behavior from him.

THE FUTURE OF THE 3D PRINTING CULTURE

I think, like me, he had to make a choice. I had made the choice to quit my well-paid manager's job and fully commit to my business. As he later half-joked, 'Being a wannabe entrepreneur and teacher is bloody hard.' Well, to me, he just had to make the choice. He couldn't do both and succeed.

I thought I might have to get a temp job. And thinking back, I should have gotten a bigger loan and not been too optimistic at the start of the business. Thoughts turned to seeking grants, especially for start-up funding for black businesses.

I had to start making hard financial decisions. I had to cancel direct debits and start saving money. I had a bit of a lifeline from a consolidation company offering debt management. While I couldn't get a consolidation loan, they could handle and reduce my repayments to loan and credit companies. That would save me some money each month, which took some pressure off repayments, and I could get some food in, but it still meant having to sell the course.

If I didn't want to have to get a temp or part-time job straight away, I was going to have to redouble my efforts at selling the course, affiliate products, and my products. And fast! I was still selling personal items on eBay and Facebook to make ends meet. I did try picking up affiliate 'contractors' on Upwork and Freelancer who could do the selling for me once I had gained the client. It would have been more of a partnership splitting the profits. I had some offers, but they didn't pan out.

Why does everyone assume, especially those giving advice on investment, that everyone has a bank of mom and dad or a circle of friends and family willing to throw money at you? My parents are pensioners. I wanted to be sending them money, not vice versa. And quite frankly I like to keep business and friends separate. Support is fine, but money and friendship don't always mix as I've found out.

Things were getting desperate by the last full week of the month. I was running out of food and supplies (lol, I sound like I'm on an Antarctic expedition!) and nothing was selling on eBay, Facebook, or other marketplaces. I had to apply for universal credit eventually succeeding with the first payment in February. No job offers were coming through, even immediate start jobs, though I was signed up to a myriad of online surveys which took ages to get any payment back.

February 2022

After that, it was a fast slide downward from a failed pivot to create a 3D printing course, which ultimately hinged on relying on someone who didn't want to leave their job, take a chance, and become an entrepreneur as much as they said they wanted to be. In hindsight, I should have realized that it was me making all the plans, trying to steer the meetings, and striving for new ideas. He was a sounding board and nothing else. But when it came down to his efforts to create a course, it was cut short by a family emergency from which he never returned. After that, we stopped communicating.

By now, I had invested over £9000.00 in creating a new type of 3D printing company and still returned less than £100.00. Not good. I had hoped the Academy of 3D with its accredited courses would have saved me. I looked for other freelancers and engineers to take on the course presentation, but as I didn't have anything to pay them upfront, many declined, not thinking they'd be paid from the profits of course sales. I tried to entice contenders to do the 3D printing course for free until we got paid for the course on the online platform.

One engineer who was keen to do so then contacted me two weeks later to say he hadn't read the syllabus properly and didn't even own a 3D printer so he couldn't do the practical lessons. Others wanted half upfront, which I tried to meet. And, at least I wasn't the only one with business partner issues as another candidate wanted to do the course but had to wait on his partner's input who in the end decided to decline.

For me, that left me bereft of the time and money to do anything else and I sunk from there. 3djacent was practically dead. I made my last payments for Printful, Shopify, Xero in January and slowly lost the websites, shops, and email access due to non-payment of bills. The last act was to pay the next Confirmation Statement for Companies House. It was heartbreaking. And of course, I felt like a failure.

Then in early February, my landlord told me that I would have to move out in May as he was converting the flats back into a single house for his family to live in. So not only did I have mounting debts, no job, and job offers, but I was soon to be without a flat.

March 2022

I was now getting universal credit, which kept a roof over my head and paid a few bills. I had applied for work, successfully getting three temp/freelance jobs; one of which was as a film/TV extra, another as a ghostwriter, and the last going back to my roots in property management though as a temp concierge. With the company gone, these jobs and universal credit were my main income for the year and beyond.

October 2022

I was done. The company was done! During the year, I thought if only I could get something going, I'd get the company back up and running. I looked up my website on GoDaddy only to find it was now with a broker, so I'd have to pay to get it back. So, I decided to make the company dormant. I had already stopped trading in February 2022 and nothing was coming in. The corporation tax was upcoming and I had to get someone to do them for me. As it was due at the end of the month, I had to contact both Companies House and the HMRC for an extension with the former granting a 3-month extension and the latter stating I would receive a penalty notice for late filing.

I was still at the same flat as I couldn't move anywhere. No landlord or agent would take me while I was on universal credit (which was discriminatory and disheartening as my housing benefit would have guaranteed rent payment). And my landlord, who was understanding, had started loft conversion works on the flat. It was a frustrating period.

I also got rid of the 3djacent Twitter and Instagram accounts, eventually canceling them altogether. I kept the Facebook and LinkedIn accounts so I could touch base with 3D printers should 3djacent make a comeback.

I also heard from my website developers who were looking to see how their website and my company were performing. While it was hard delivering the news to them that 3djacent had failed and I had lost the website through non-payment, they were supportive, suggesting I could sell the company.

It was at this time I finalized the idea for my two TV 3D printing shows. I researched independent producers who would accept unsolicited submissions. There weren't many. So, as I had worked unpaid at an independent TV production company over a decade ago, I reached out to them to see if they would accept my ideas. I didn't send them the ideas, just told them it was about a new type of engineering show.

One of the directors, David Dugan from Windfall Films responded that even they were:

> "reluctant to take unsolicited ideas in case they have something similar in development – and then are accused of stealing that idea. It's a weird Catch 22 I know…" though he then stated there was still: "room in the TV market for a new reality format in the engineering area. There hasn't really been anything significant since Scrapheap Challenge/Junk Wars."

He asked for a chat during November. My hopes were high.

November 2022

Even with no business to run I still had to pay my corporation tax and accounts for Companies House. I was lucky enough to get some money in to hire an accountancy company who had sent me a letter through the post — something that I would have considered junkmail a few years ago. And as I had not traded since February, I was able to apply for their lower payment rate.

I had not heard back from the TV director, so I emailed him mid-November, though I received no response.

And then I got my Section 21 notice from the landlord to vacate the property within 2 months. Things were getting tense again in the run-up to another Christmas. I started looking for cheaper non-private accommodation options and also turned to the council for help.

December 2022

There was some good news in that my accounts and corporation tax were filed by the accountancy agency.

However, there was no word from the TV producer after two more emails and I offered to hear from them after the holidays.

Writing gave me some solace and I continued trying to contact companies to solicit my ideas and receive some feedback on their industry, such as my communications with Stanley Black + Decker and Ikea on the furniture ideas.

Sean contacted me to see how I was doing and continued to give me words of encouragement, but I had no positive news for him. I told him how the year had gone and he said if I'm back in business and have something to sell, he'll deal with the leads, convert them, and I'd make commission — quickest way to get back on my feet, he said. I replied I would look into options and get back to him.

January 2023

I sent another email to the TV director. Without a response, I decided I would send them the TV show ideas after I had published them in this book. That way the idea couldn't be stolen wholesale and I could inspire a 3D printing show on TV.

I still had a small savings pot left through successful film/TV roles, three ghostwritten books, and the continued temp concierge jobs. Though work was becoming scarce. It was time to start looking for something more sustainable.

The Section 21 notice had expired. I had not found a place even with the council's help and the landlord talked about commencing court action.

February 2023

Another year had passed and another Companies House Confirmation Statement was paid for and filed. At least 3djacent files were up to date, though it would have been nice to have access to the website and shops and access my next steps for action. But, in the end, I decided not to strike 3djacent from Companies House as I stubbornly refused to let it die. I just wouldn't be trading until I had a new plan to rescue 3djacent or sell it.

I sent my ideas to XPRIZE and also contacted Climeworks and the University of Surrey regarding my ideas on the carbon-capture-to-carbon-3d-printing process.

March 2023

I started agent hunting for my books; sci-fi, non-fiction, and this current book. While I had a few responses, I wasn't successful with this, but I wasn't too bothered. Over the years, I had tried the traditional publishing route, but preferred self-publishing. I felt, however, that this book was different and could attract a traditional publisher. In the end, I may self-publish and then submit to agencies that accept previously published works. As I told other author friends before, it's always better to publish for yourself rather than waiting on others to do so for you, or you'll never get your work and ideas out there. And the way traditional publishing is now, you might as well be a self-publisher for all it's worth.

April 2023

I was disappointed that my letters to Molyworks, Stanley Black + Decker, Ikea, and others were not replied to. Also, the TV production company was silent. In some ways, I can understand their reticence seeing as I'm not a famous author or celebrity or they didn't want to be involved in intellectual property issues.

But such issues could have been worked out even with a simple reply. Ideas are free. It's the application that holds the value and I was just happy to spread knowledge and learn.

On a good news front, my landlord could not go ahead with the second part of the eviction process due to costs and extra maintenance work to the flat, he would incur in having to go through the courts. So, now it was up to me to get myself out of the flat.

I decided it was time to reskill. I didn't want to return to property management yet. However, my writing and the TV/film extras work was not bringing in enough. I chose to take on software engineering, a free government-funded 3-month bootcamp course with job interviews at the end. I had never engaged in this before and wasn't sure what to expect. While not exactly math (which I hated as a kid) I reckoned as a writer I would have to view Python coding as a new form of creative writing albeit with jumbled-up letters and numbers. But, it was time to take a new step. Maybe I would even use my skills in the 3D printing industry. This idea didn't last long, as I didn't complete the first part of the course, the Microsoft Certified Azure Fundamentals course. My brain ouldn't kick in and no matter how much I studied, the materials wouldn't :ick. If I had more time rather than two weeks, I probably could have done , but I found the pace of the course daunting. I still have the option to go it lone and pay for the exam myself. But my mind was telling me to stick with 1y creative side.

astly, up until a few years ago, I and fellow author friends would attend the nnual London Book Fair in Olympia, west London. While I did not attend the vent this year, I was able to meet up with my friends at an after-party, swapping 1e usual news of our successes or author journeys. But such events are also etworking opportunities with new people, whether other authors, publishers, gents, or companies. One such company was Book Award Pro. I spoke with s CEO, Jay Jacobson. I liked the premise of marketing and promoting books 1rough custom award/review matching targeting my most valuable pportunities even for a book in a pre-published state. I corresponded with im and the author success manager, Nour Youssef, on setting up my account or a free-month's trial. What could be better than winning an award for my ook, possibly coupled with an XPRIZE winning opportunity? So, I took the lunge and entered this book for an award. And who knows, maybe it would ttract an agent.

lay 2023

was delighted then, when mid-month, I was contacted by Book Award Pro at my book had been matched with a possible award by The Literary Titan in e technology category. So, I was in the running for an award and possibly an PRIZE entrant.

p until then, life had been ticking along, with the temp concierge work keeping e afloat. The ghostwriting had dried up and filming work was intermittent. ut I felt writing this was a beacon of hope for me.

June 2023

As noted in Chapter Eleven, this book did not make it to the next phase of the XPRIZE competition, but I was extremely proud to have made it that far. Hopefully, there are other opportunities in the future. However, as with the possible award, I had posted my book's journey on Facebook and LinkedIn. And I had worked on a cover page with clip art on an isometric background. I wanted something simple yet told the story of the book. Hopefully, that was translated on the cover. My plan was to publish as an ebook first and when I had some funds, to print in hardback and/or paperback.

The next phase on my journey was my first-ever pre-pub review by The Literary Titan. It was very positive and welcomed my ideas about the future of 3D printing. I then had to complete an author's interview with them, which was published near the end of June. I was still working feverishly to edit and publish the book. And I was still in with a chance to win the Literary Titan Book Award in the technology category in July.

July 2023

In the 1st edition of this book, the below was the addendum. However, more has happened since. I had finished the final edits on my book and planning my search for beta readers. It was 10pm and I decided to get a late shave as the landlord had been in doing works in the flat and I hadn't had the time earlier. Once I'd lathered up my face with foam, I decided to check my emails. In the junk folder, which was strange as such emails were previously in my inbox, was a email from Thomas Anderson, Editor In Chief at Literary Titan. The top line was Congratulations with a trophy icon.

"Oh my God, have I've just won an award for my book?"

I opened the email, and yes, incredulously, I had won Literary Titan's silver award in the technology category.

read the whole email. I had a little cry. Everything I had done in the past 2 years came flooding back to me. I told my family, thanked Literary Titan, then had my shave.

then had to think about more edits, hitting social media (at least Facebook and LinkedIn, sort out my Amazon and Goodreads author pages) to spread the news, starting a new Kindle entry, and perhaps going back to the TV producer. Would an award open doors for me? I hoped to find out.

November 2023

As I had noted above, I had lost my GoDaddy domain and Shopify site with my shirts and other merchandise. I contacted them to see if anything was archived but there was nothing. I then went back to Webefy who had built my website but as I had not retained them for site maintenance, they had not kept a backup, so 3djacent had no online presence. Interestingly, when speaking with GoDaddy, I was told that the broker who had my domain was only a middleman and the owner could even refuse to sell the domain to me, so it looks like I would have to create a new domain to avoid any silly bidding wars.

December 2023

I was late. I am the director of my company and trading or non-trading, money or no money, I am still responsible for filing my Corporation Tax and company accounts for the end of October. I was aware of the lateness, but it took me a while to come up with the funds to pay the accountancy company. By the beginning of December, I had them filed, but far too late to avoid a penalty notice. Lesson learned.

January/February 2024

Four years on, and in a way, I am looking forward to 2024. I'll be back on my feet and feeling encouraged to re-engage with 3djacent. In what form, I'm not sure yet. However, writing this book and reading other 3D printing stories online has left me feeling more positive that the 3D printing culture is on the way.

It's been quite a journey from a 4-day 3D printing course, to thinking about the trajectory of the industry, and gaining some validation with an award for my ideas almost 2 years after leaving full-time employment to try and set up my own company. And I thank you all who have personally shared my journey or who are now reading this book. I hope it inspires you on your own 3D printing journey.

The Lessons of 3djacent

During my trials and tribulations, I learned quite a lot about myself and business in general. I have tried to coalesce them into some general learning points and tips.

Lessons Learned

- Don't do too much the wrong way. Do one good thing at a time the right way.
- Don't deny problems. Meet any issues head-on by finding the root cause and try to resolve them rather than delay and defer. I think I procrastinated too much waiting to see how things panned out and hoping for the best.
- Be positive and have self-belief, but not to the point of over-exuberance or being too confident.

- Look after your physical and mental health. I'm not one to suffer from depression or anxiety, but even at the lowest point, I kept myself going physically and mentally. I did suffer more physically from sitting most of the day whether at home working or not, or when at work still sitting. Walking, running, exercising, and stretching were important to me. Just getting some sun and breathing fresh air did a world of good.

Criticism: don't ignore it, but take constructive criticism when offered. However, don't keep taking it personally and conversely don't take every critical remark by someone else as gospel. Defend your ideas, reevaluate them, and check your own biases for and against your ideas and business decisions.

Incentivize yourself and hone your business instincts to what works and what doesn't. I didn't have any formal business training and though I received lots of advice from the entrepreneurs' Meetup, not all of it was relevant or adaptable to my situation. I had to pick and choose and go with the flow sometimes.

Stick to your strengths. I am not a salesman, marketer, designer, or engineer. I'm an ideas person, the Chief Creative Officer. Sometimes my ideas just spill out on the page, unfiltered, like on my website which can be overwhelming. I needed a strong partner to check or at least moderate my excesses (especially spending just on marketing). But lucky for you, you get to read about all my ideas here.

Confront potential business partners early. Challenge them and their commitment to the business. I certainly prevaricated with this. I didn't want conflict and trusted too much in their words and not their actions. I knew what I wanted from my business. Danny didn't. It wasn't a clash in vision, more of a clash in ambition and personalities.

Pivot the business to a more reliable profit maker to steady the ship and then re-invest in the original idea or move on. I looked for options too late, choosing to stick with the affiliate plan which was not working at all.

- Ensure you have a secured funding pot or runway money to survive the first few months of little to no trading or profit. In hindsight maybe I should have been more ambitious and secured an initial loan for £20k rather than £7.5k, followed by a second for £5k. I thought I could survive 3 months until I made a profit but that did not happen and I had to take out a third loan. My financial approach was scattershot and now I have repayments to 3 creditors and nothing to show for it.

3djacent Highs and Lows

Highs

- Paris. Launching 3djacent in Paris was a big deal for me visiting another country on a business venture. It made me feel legit. And the food… ooh lala…
- Creating my own bespoke website. Seeing this come together from my drawings and texts, built frame by frame and page by page was inspiring.
- Creating a collection of branded t-shirts, books, marketing materials, and office supplies.
- Having the ability to envision the cultural future of 3D printing and writing this book to spread the word.

Lows

- Loans. As lamented above, I should have had a better grasp on the financials and been smarter in conserving money.
- Not being able to utilize any of my affiliates effectively. Either I had to learn how to market them and/or choose the right one.

- People. While I met some great people in real life or online, there were times I had to rely on others for my business to grow. Either I have to become more self-sufficient or learn to effectively delegate with the right person for the right job.

Plans

- As I'm not trading, I had planned to make 3djacent dormant, but that, too, costs money to enact, so I will leave it in a non-trading state for now.
- Since rescuing the website/domain is no longer an option, I could invest in a new website, new domain, and/or Shopify store. Even though I've lost my Shopify and Printful accounts with my t-shirt merchandise, I still have copies of the advertising pictures and original designs, so I could always reconstruct the t-shirt line and restart the shop in the future. Any takers to join me?
- Renew and Pivot? If there is enough incentive and interest, I may start trading again with a modified version of the company, perhaps with courses.
- Renew and sell? Maybe the best course of action will be to secure a new domain, reinstate the shop and sell the company.

he above plans were from pre-2024. So, I had a decision to make. 3djacent olutions Ltd had existed for just over 4 years, was active for around 18 months, id now is closed. There may be hope down the line for a revamp or do-over, but least my ideas for the 3D printing culture continue within this book.

nd that is the story of 3djacent. I knew what I wanted but ultimately not ow to put it out there. There was no real plan, just my ideas splashed across a ebsite. As remarked above, I was trying to do too much the wrong way rather an doing one good thing at a time the right way.

I somewhat feel like a delinquent skiving off from traditional work for over two years. I was financially poorer but happier and richer in some sense, within myself. But I knew I had wanted to try something different, to challenge myself at this stage in my life, to think and work flexibly. If I hadn't done so, then I would never have known. And it would have been a great regret. In some ways, I achieved my dream, but in others, I failed. But it's the journey that counts.

And the discourse on the future of the 3D printing culture journey continues. Technology changes, but how will our cultural attitudes towards 3D printing evolve? I'll still fight for it. As I have stated throughout, I think I'm a little ahead of time. But now that you know, I hope some of you will help carry that torch and shine a light on the bright future of the 3D printing culture.

My 3D Printing TV Program Ideas

3D Print Wars

remise: 3D printing meets Bake Off meets Grand Designs.

Channel: Channel 5/Blaze/Quest.

JSP: Reality TV meets 3D Printing.

alent:

x hosts (one presenter like Tim Shaw, Rob Bell, Kate Bellingham, or ran Scott and one 3D print design engineer similar to Youtube stars James ruton, Richard Horne, or Simone Fontana) with a weekly guest known as The client'.

ormat:

ach week in the 1 x hour programme, 4 teams of 3D printers compete to omplete a task set by the mystery Client. The identity of the Client is held ack at first to inject some surprise for the teams and also to prevent bias, iving the teams a specific project brief for a blind test to complete (Think of he cooking test laid out by a visiting guest in MasterChef). Whether the 3D roduct is produced through plastic, biomaterial, metal, resin, ceramics, ood, etc., the teams have a set deadline.

here will be 2 rounds per episode. For the first 'blind test' round, the Client ill remain unknown to the teams. The hosts will then walk and talk mongst the teams, getting to know them, what they do, and their 3D print istories. We will watch the teams as they explain their technical set-up and tart their 3D modelling designs. Once the designs are set, the hosts will hen start the competition with: "Let's Print!"

he teams will press the printer buttons and begin printing their version of he Client's brief. The Client will then be revealed and after their inspection of he products will choose the 2 teams who best personify the brief thus liminating the other teams.

The final 2 teams will then go head-to-head to accomplish the Client's 3D printing goal whether to build a superhero costume, or a house, create a new car component or medical device, produce a 3D-printed gourmet meal, or even create a new 3D printer for the mass market.

As prints can last hours or even days, time-lapse will be used to speed up the filming and to make sure the varied prints finish at the same time for the programme. Such a sequence or montage will also highlight any production issues, such as printing faults, for the teams and the Client to judge.

Upon print completion and announcement of the winner, the Client will reveal the prize, whether monetary, a job or contract with the Client, or 3D printing equipment, advertising, or exhibitions.

The programme location will be studio based and the equipment can be brought in by the contestants, whether a desktop/household printer or a commercial/industrial printer chosen by the Client depending on the work. The eliminated projects can be recycled after production.

Recycling:

Recycling will be a feature of the show not just plastic waste, but also metals and ceramics which can also be recycled through professional services. The safety and environmental impact of 3D printing will be emphasised.

Audience:

3D Print Wars will be a programme for all ages, enticing young adults interested in the 3D printing industry who want to learn and earn through 3D printing at home, and industry experts keen to see their profession through a different media and platform.

So, following in the footsteps of reality TV cooking, sewing, painting, job hunting/talent spotting, and surviving, let's open up the competitive spirit of 3D printing. Let's print!

End.

Make It For Me

Premise: 3D printing meets Grand Designs meets The Fantastical Factory of Curious Craft.

Channel: Channel 5/Blaze/Quest.

USP: Design & Craft with 3D Printing.

Talent:

Six hosts and a selected team of 3D printers (one presenter like Tim Shaw, Rob Bell, Kate Bellingham, or Fran Scott and one 3D print design engineer similar to Youtube stars James Bruton, Richard Horne, or Simone Fontana). The team will be drafted in like a 'Mission Impossible' team each with particular skills.

Format:

Each week in the 1 x hour programme, a client, whether commercial or residential, asks the selected TV team of 3D printers to create a specific or customised project or object for them whether a statue, a superhero costume, a car chassis, or medical equipment. The products they can use will range from plastic, biomaterial, metal, resin, ceramics, food, wood, etc.

The object can be an original piece, bespoke, or a replacement for a lost or broken object where the expense to replace would be far costlier than to 3D print it.

It may be an exhibition, objet d'art, or a personal piece. For projects, the client may even want a house or community centre built for them. Fancy a 3D-printed steak or burger, chocolate cake, or a pasta castle? Let the team create it.

We will watch as the team designs the project/object using software and programmes the printer. As prints can last hours or even days, time-lapse will be used to speed up the filming. Such a sequence or montage will highlight the build process and any production issues, such as printing faults, which the team will correct before the final presentation.

Upon print completion, the team will present the client with their build. The client may be blindfolded before the reveal or covers drawn off the creation for the surprise effect.

The programme location will be studio-based using desktop/household or a commercial/industrial printers for most of the builds, unless an on-site build is required, for instance for a house or other external builds. Smaller, portable builds can be transported to the client. Any waste or eliminated project/object designs can be recycled after production.

Recycling

Recycling will be a feature of the show not just plastic waste, but also metals and ceramics which can also be recycled through professional services. The safety and environmental impact of 3D printing will be emphasised.

Audience:

Make it For Me will be a programme for all ages, enticing young adults interested in the 3D printing industry who want to learn and earn through 3D printing at home, and industry experts keen to see their profession through a different media and platform.

So, following in the footsteps of reality TV cooking, sewing, painting, job hunting/talent spotting, and surviving, let's open up the competitive spirit of 3D printing.

Let's print!

End.

Printer Winner

Premise:3D printing meets Grand Designs meets Master Chef.

Channel: Channel 5/Blaze/Quest.

USP: Design & Craft Competition with 3D Printing.

Talent:

2 x hosts (one presenter like Tim Shaw, Rob Bell, Kate Bellingham, or Fran Scott and one 3D print design engineer similar to YouTube stars James Bruton, Richard Horne, or Simone Fontana).

Format: 1 x hour programme.

Each week 4-6 individuals will compete in head-to-head knock-out stages of design and print challenges. They will be given a brief by the hosts and then have an hour to design and slice/code the item. The products they can use will range from plastic, biomaterial, metal, resin, ceramics, food, wood, etc.

We will watch as the team designs the project/object using software and programmes for the printer. The print times may vary, but the maximum time will be twelve hours. To facilitate this on-screen, action will be split between normal and time-lapse montages. The contestants will also complete any post-processing (such as polishing, deburring, support removal, etc.) within the allotted time.

During the design phase, the audience will get to know each of the contestants as the hosts ask them questions regarding their background in 3D printing and their tasks for that episode. This is a programme about the personalities, the mavericks, the budding entrepreneurs, and the entertainers of the 3D printer world. Over the weeks, the series will culminate in quarter- finals, semi-finals, and the grand finale to crown the annual 3D printer champion.

The programme location will be studio-based using desktop/household printers. The programme also has the potential to 'Go Large' with **Monster Printer Winner** where contestants get to compete using commercial/ industrial 3D printers for larger prints.

Recycling

Recycling will be a feature of the show. Any waste or eliminated project/object designs can be recycled after production, not just plastic waste, but also metals and ceramics which can also be recycled through professional services. The safety and environmental impact of 3D printing will be emphasised.

Audience:

Printer Winner will be a programme for all ages, enticing young adults interested in the 3D printing industry who want to learn and earn through 3D printing at home, and industry experts keen to see their profession through a different media and platform.

So, following in the footsteps of reality TV cooking, sewing, painting, j hunting/talent spotting, and surviving, let's open up the competitive spirit 3D printing.

Let's print!

End.

You

have your future in your hands

Think on it, design it, make it.

Your 3D printing lifestyle resides

within these pages in your ideas and projects.

Exceed your expectations

as the 3D printing horizon expands.

Let's get started...

REFERENCES

3D Printing Industry. (2016, April 17). World tour: The best physical stores to buy a 3D printer. 3D Printing Industry. https://3dprintingindustry.com/news/world-tour-physical-stores-to-buy-a-3d-printer-76506/

3D Printlife Algix ALGA Algae Based PLA 3D Printer Filament. (n.d.). www.3dprintlife.com. https://www.3dprintlife.com/alga

Adamson, D. (2022, May 31). *Mining Noise Control: Why Crypto Mining Rigs Make so Much Noise and How to Combat it?* www.coinIdeology.com https://www.coinideology.com/mining-noise-control

Ajadi, L. (2020, June 10). *Can you 3D Print with glass? | Solid Print3D.* www.solidprint3d.co.uk. https://www.solidprint3d.co.uk/can-you-3d-print-with-glass/

American Academy of Audiology. (2010). *Levels of Noise In Decibels (dB).* https://audiology-web.s3.amazonaws.com/migrated/NoiseChart Poster%208.5x11.pdf_5399b289427535.32730330.pdf

B4Plastics | Redesigning tomorrow's plastics. Today. (n.d.). B4Plastics. https://b4plastics.com/

Baldwin, J. (2021, July 21). *Countries that are leading the way in 3D Printing.* GrabCAD Blog. https://blog.grabcad.com/blog/2021/07/21/countries-that-are-leading-the-way-in-3d-printing/

Bello, A.-R. O. (2023, May 12). *Breakthrough in 3D printing: elastic conductors for stretchable electronics.* www.Interestingengineering.com. https://interestingengineering.com/innovation/3d-printing-elastic-conductors-for-stretchable-electronics

brettlarenatkins. (2021, June 21). *What does THRINT mean?* www.definitions.net. https://www.definitions.net/definition/THRINT

Brittle, C. (2023, March 3). *F1 2023 commercial guide: Every team, every sponsor, all the major TV deals.* SportsPro. https://www.sportspromedia.com/analysis/f1-2023-commercial-guide-teams-sponsors-tv-rights-deals/?zephr_sso_ott=5AhDG8

Brooks, M. (2022, June 23). *How Much Power Does a 3D Printer Use? [Electricity Costs].* M3dzone.com. https://m3dzone.com/3d-printer-power-usage/

Campelli, M. (2020, January 16). *"We're not here to educate fans – we make sustainability fun" – Sustainability Report.* https://Sustainabilityreport.com/ https://sustainabilityreport.com/2020/01/16/were-not-here-to-educate-fans-we-make-sustainability-fun/

Carter, C. (2022, October 3). *Bee-like aerial 3D printing.* www.superinnovators.com. https://superinnovators.com/2022/10/bee-like-aerial-3d-printing/

Connelly, L. (2021, May 20). *Reprint Ceramics: Sustainability paired with 3D printing and parametric design.* Material Source. https://www.materialsource.co.uk/reprint-ceramics-sustainability-paired-with-3d-printing-and-parametric-design/

Diamandis, P. H., & Kotler, S. (2014). *Abundance : the future is better than you think*. Free Press.

Diamandis, P. H., & Kotler, S. (2015). *Bold : how to go big, create wealth and impact the world*. Simon & Schuster.

Diamandis, P. H., & Kotler, S. (2020). *The future is faster than you think : how converging technologies are transforming business, industries, and our lives*. Simon & Schuster Paperbacks.

Easy Crypto Hunter. (n.d.). *Do Mining Rigs Give Off Heat?* Easy Crypto Hunter. Retrieved March 15, 2023, from https://www.easycryptohunter.co.uk/what-is-cryptocurrency-mining/do-mining-rigs-give-off-heat/

Frankopan, P. (2023). *The Earth Transformed*. Knopf.

Gershenfeld, N. (2000). *When things start to think*. Holt.

Gokce Bahcegul, E., Bahcegul, E., & Ozkan, N. (2022). 3D printing of crude lignocellulosic biomass extracts containing hemicellulose and lignin. *Industrial Crops and Products*, *186*, 115234. https://doi.org/10.1016/j.indcrop.2022.115234

Goodbaum , B. (2023, July 17). America Makes and ANSI publish standardization roadmap for additive manufacturing version 3.0 . www.ansi.org. https://www.ansi.org/standards-news/all-news/2023/07/7-17-23-america-makes-and-ansi-publish-standardization-roadmap-for-additive-manufacturing-version-3

GREENFILL3D - 3D printing with eco filaments. (n.d.). GREENFILL3D. Retrieved July 1, 2023, from https://greenfill3d.com/

Gururaj, T. (2023, July 5). *Scientists create highly conductive metallic gel for 3D printing at room temperature.* www.Interestingengineering. com. https://interestingengineering.com/innovation/ conductive-metallic-gel-3d-printing-room-temperature?utm source=Facebook&utm medium=content&utm campaign=organic&utm content=Jul05&fbclid=IwAR2I7yEjxd4 9f-QLfdttDcs3ko FoCxz3p27BafiUGkWVKXavs-Bi3x961M

Hardman, B. (2021, December 23). *Ceramic Recycling: Is Ceramic Recyclable & How To Dispose?* www.tinyecohomelife.com. https://www. tinyecohomelife.com/ceramic-recycling

Hazzard, T. (2016, May 31). *3D Print Power Consumption – How Much Power Does a 3D Printer Use?* 3D Start Point; 3D StartPoint. https://3dstartpoint.com/3d-print-power-consumption-how-much- power-does-a-3d-printer-use/

Hendrixson, S. (2020, July 27). *Is Recycled Metal Scrap the Future Feedstock of Choice for Metal 3D Printing?* www.additivemanufacturing.media. https://www.additivemanufacturing.media/articles/is-recycled-metal- scrap-the-future-feedstock-of-choice-for-metal-3d-printing

Hilding, T. (2021, May 17). *New technology converts waste plastics to jet fuel in an hour.* ScienceDaily. https://www.sciencedaily.com/ releases/2021/05/210517124937.htm

IEA. (2019, September). *Putting CO2 to Use – Analysis*. IEA. https://www.iea.org/reports/putting-co2-to-use

ISO Technical Committee. (2022, March 31). ISO - Smart manufacturing - New generation plastics- Additive manufacturing. ISO. https://www.iso.org/foresight/smart-manufacturing.html

Jezos, B., & bayeslord. (2022, July 10). *Notes on e/acc principles and tenets*. Beff's Newsletter. https://beff.substack.com/p/notes-on-eacc-principles-and-tenets

James Lovelock: You Ask The Questions. (2006). *Physics Today*. https://doi.org/10.1063/pt.5.020348 (Accessed 16.06.08. Thread no longer active)

Jowit, J. (2008, June 7). Councils to store nuclear waste in return for cash. *The Observer*. http://www.guardian.co.uk/environment/2008/jun/08/nuclearpower.waste?gusrc=rss&feed=society

Kang, X., Jia, S., Xu, R., Liu, S., Peng, J., Yu, H., & Zhou, X. (2021). Highly efficient pyroelectric generator for waste heat recovery without auxiliary device. *Nano Energy, 88*, 106245. https://doi.org/10.1016/j.nanoen.2021.106245

Kasat, R. (2023, October 6). 3D Systems' EXT Titan Pellet Printers Help Model No. Redefine Sustainable High-End Furniture | 3D Systems. www.3dsystems.com. https://www.3dsystems.com/customer-stories/ext-titan-pellet-printers-help-model-no-redefine-sustainable-high-end-furniture

M, M. (2023, October 18). Dubai Launches World's First Certification Program for the 3D Printing Construction Sector. www.3Dnatives. com. https://www.3dnatives.com/en/dubai-launches-first-certification-program-for-3d-printing-construction-181020236/#

Malayil, J. (2024, January 10). Post-phone era? Tiny "rabbit" AI device ditches apps for real-world magic. www.interestingengineering.com. https://interestingengineering.com/innovation/tiny-ai-device-ditches-apps?fbclid=IwAR2RNRXYvuV7Uv4NJ-8MJ8eU53qJw0xMiJtW2t5FGWJPzsQDugtKZjudJgo

Minkel, J.R. (2002). *The Meaning of Life*. New Scientist. 176 (2363), 30-33.

Mulhall, D. (2002). *Our Molecular Future*. Prometheus Books.

Nanyang Technological University. (2022, April 26). *Scientists use recycled glass waste as sand replacement in 3D printing*. ScienceDaily.com https://www.sciencedaily.com/releases/2022/04/220426101737.htm

NoiseBarrierWalls.com. (n.d.). *Effective Bitcoin Mining Noise Reduction Solutions*. Noise Barrier Walls. Retrieved March 15, 2023, from https://noisebarrierwalls.com/applications/crypto-and-bitcoin-mining-noise-reduction/#:~:text=A%20single%20bitcoin%20mining%20rig

O'Neal, B. (2020, June 27). *Researchers Review 3D Printing with Biomass-Derived Composites*. www.3DPrint.com | the Voice of 3D Printing / Additive Manufacturing. https://3dprint.com/269253/us-researchers-review-latest-3d-printing-biomass-derived-composites/

Obudho, B. (2018, December 21). *The 3D Printer Filament Recycler's Guide*. All3DP; All3DP.com https://all3dp.com/2/the-3d-printer-filament-recycler-s-guide/

P, M. (n.d.). *Is 3D Printing in Marketing a New Advertising Tool?* 3Dnatives. Retrieved July 13, 2023, from https://www-3dnatives-com.cdn. ampproject.org/c/s/www.3dnatives.com/en/is-3d-printing-in-marketing-a-new-advertising-tool-100720234/amp/

P, M. (2023). *New Sinter-Free Process Can Make Nanometer-Sized, 3D Printed Glass Structures.* 3Dnatives.com. https://www-3dnatives-com. cdn.ampproject.org/c/s/www.3dnatives.com/en/new-sinter-free-process-nanometer-3d-printed-glass-200620234/amp/

P, M. (2023b, September 18). *Tesla Turns to 3D Printing to "Reinvent Carmaking."* 3Dnatives. https://www.3dnatives.com/en/tesla-turns-to-3d-printing-to-reinvent-carmaking-190920234/#

Papadopoulos, L. (2022a, June 29). *World's largest direct air carbon capture facility begins construction in Iceland.* Interestingengineering.com. https://interestingengineering.com/innovation/worlds-largest-direct-air-carbon-capture-facility

Papadopoulos, L. (2022b, November 12). *New CO2-capturing technology could help combat the climate crisis.* Interestingengineering.com. https:// interestingengineering.com/science/new-technology-captures-co2

Pigott, P. (2023, October 16). Climate crisis: Coca-Cola trial to make bottle tops from CO2 emissions. BBC News. https://www.bbc.co.uk/news/uk-wales-67060151

rayjacent. (2023a, January 17). *Urban Dictionary: Thrab.* Urban Dictionary. https://www.urbandictionary.com/define.php?term=Thrab

rayjacent. (2023b, January 17). *What does thrab mean?* www.definitions.net. https://www.definitions.net/definition/thrab

rayjacent. (2023c, July 4). *Urban Dictionary: 3Der*. Urban Dictionary. https://www.urbandictionary.com/define.php?term=3Der

rayjacent. (2023d, July 4). *What does 3Der mean?* www.definitions.net. https://www.definitions.net/definition/3Der

rayjacent. (2023e, August 20). *Urban Dictionary: Metaprinting*. Urban Dictionary. https://www.urbandictionary.com/define. php?term=Metaprinting

rayjacent. (2023f, August 20). *Urban Dictionary: Thratter*. Urban Dictionary. https://www.urbandictionary.com/define.php?term=Thratter

rayjacent. (2023g, August 20). *Urban Dictionary: Throdel*. Urban Dictionary. https://www.urbandictionary.com/define.php?term=Throdel

rayjacent. (2023h, August 20). *What does Metaprinting mean?* www. definitions.net. https://www.definitions.net/definition/Metaprinting

rayjacent. (2023i, August 20). *What does Thratter mean?* www.definitions. net. https://www.definitions.net/definition/Thratter

rayjacent. (2023j, August 20). *What does Throdel mean?* www.definitions.net. https://www.definitions.net/definition/Throdel

Rob. (n.d.). *Do 3D Printers Use a Lot of Power? (The Numbers Inside)*. 3D Printscape. https://3dprintscape.com/do-3d-printers-use-a-lot-of-power/

S, C. (2023, March 2). *How is 3D Printing Used in the Nuclear Power Sector?* www.3Dnatives.com https://www.3dnatives.com/en/3d-printing- nuclear-power-sector-020320235/

Samarthya Bhagia: 3D printing with biomass-based materials | *ORNL.* (n.d.). www.ornl.gov. Retrieved July 1, 2023, from https://www.ornl.gov/content/samarthya-bhagia-3d-printing-biomass-based-materials

Schwaar, C. (2022, September 28). *Eco Resin Advancements - 3D Printing with Recycled Resin.* All3dp.com. https://all3dp.com/4/3d-printing-with-recycled-resin/

Simplify3D. (n.d.). *Ultimate Materials Guide - 3D Printing with Carbon Fiber.* Simplify3D.com. https://www.simplify3d.com/resources/materials-guide/carbon-fiber-filled/

Sky Diamond. (n.d.). *Home.* Skydiamond.com. https://skydiamond.com/

Souty. (2021, June 26). *Urban Dictionary: thrint.* Urban Dictionary. https://www.urbandictionary.com/define.php?term=thrint

Stevenson, K. (2018, May 24). *Resin Print Recycling Method Invented: Is The Planet Saved? «Fabbaloo.* Fabbaloo. https://www.fabbaloo.com/2018/05/resin-print-recycling-method-invented-is-the-planet-saved

Stoltz, M. (2023, September 7). *Mars Society to Launch Mars Technology Institute.* The Mars Society. https://www.marssociety.org/news/2023/09/06/mars-society-to-launch-mars-technology-institute/

Takahata , E. (2022, April 8). *3D-printing quality food for seniors*. Jstories. media. https://jstories.media/article/3d-printing-quality-food-for-seniors

Verdict. (2022, November 9). *Formula 1 adoption of 3D printing technologies*. Verdict. https://www.verdict.co.uk/formula-1-3d-printing-f1/#catfish

Voell, Z. (2022, June 2). *We Hear You: Bitcoin Mining Noise Pollution is a Solved Problem*. Bitcoin Magazine. https://bitcoinmagazine.com/business/solving-bitcoin-mining-noise-pollution

Wikipedia Contributors. (2019, November 16). *Piezoelectricity*. Wikipedia; Wikimedia Foundation. https://en.wikipedia.org/wiki/Piezoelectricity

Wikipedia Contributors. (2020, December 27). *Pyroelectricity*. Wikipedia. https://en.wikipedia.org/wiki/Pyroelectricity

Wikipedia Contributors. (2023, January 29). *Direct bonding*. Wikipedia. https://en.m.wikipedia.org/wiki/Direct_bonding#:~:text=The%20bonding%20at%20room%20temperature

Wikipedia Contributors. (2024, March 22). *Effective accelerationism*. Wikipedia. https://en.wikipedia.org/wiki/Effective_accelerationism

Williams, A. (2016, May 9). 3D Printer Prints Sound. Hackaday. https://hackaday.com/2016/05/08/3d-printer-prints-sound/

World Nuclear News. (2022, March 10). *3D printers to be used in fuel manufacture: Uranium & Fuel - World Nuclear News.* www.world-Nuclear-News.org. https://www.world-nuclear-news.org/Articles/3D-printers-to-be-used-in-fuel-manufacture

Yarckin, C. (2023, September 17). *NFL End Zone* (No. 2). Channel 5.

Young, C. (2021, October 15). *Ex-SpaceX Engineers Are Building a Cheap, Portable Nuclear Reactor.* www.interestingengineering.com https://interestingengineering.com/innovation/ex-spacex-engineers-are-building-a-cheap-portable-nuclear-reactor?fbclid=IwAR0EYZdhh8HkFIWOXxxqY558F5LoArKaJusqq7wwMLqby_d54G3tJqad_2s

Zhang, K., Chermprayong, P., Xiao, F., Tzoumanikas, D., Dams, B., Kay, S., Kocer, B. B., Burns, A., Orr, L., Choi, C., Darekar, D. D., Li, W., Hirschmann, S., Soana, V., Ngah, S. A., Sareh, S., Choubey, A., Margheri, L., Pawar, V. M., & Ball, R. J. (2022). Aerial additive manufacturing with multiple autonomous robots. *Nature, 609*(7928), 709–717. https://doi.org/10.1038/s41586-022-04988-4

Hit the book sites to write and share your reviews.

Feel free to contact me with your comments, feedback, and a chat over ideas.

I'll be at rayburke.3djacentsolutions@hotmail.com.

I look forward to hearing from you.

www.ingramcontent.com/pod-product-compliance
Lightning Source LLC
Chambersburg PA
CBHW031119020426
42333CB00012B/141